Fluid Mechanics at Interfaces 2

Fluid Mechanics at Interfaces 2

Case Studies and Instabilities

Edited by

Roger Prud'homme
Stéphane Vincent

iSTE

WILEY

First published 2022 in Great Britain and the United States by ISTE Ltd and John Wiley & Sons, Inc.

ISTE Ltd
27-37 St George's Road
London SW19 4EU
UK

www.iste.co.uk

John Wiley & Sons, Inc.
111 River Street
Hoboken, NJ 07030
USA

www.wiley.com

Library of Congress Control Number: 2021949304

British Library Cataloguing-in-Publication Data
A CIP record for this book is available from the British Library
ISBN 978-1-78630-817-7

Contents

Preface

Volume 2 of *Fluid Mechanics at Interfaces*, subtitled *Case Studies and Instabilities*, focuses on different kinds of interfaces and flow regimes in their vicinity. As with Volume 1, this book is born out of the systems and engineering e-journal, *Thermodynamics of Interfaces and Fluid Mechanics*[1], which studied thin zones separating media with different properties. These include phase separation as well as thin flames and discontinuity waves. At the macroscopic level, they are compared to material surfaces endowed with thermodynamic properties with their related constitutive laws.

We recall that Volume 1, subtitled *Methods and Diversity*, consisted of six chapters: 1. Modeling Interfaces with Fluid Phase; 2. Simulations of Turbulent Two-Phase Flows with Phase Change Using a Multifield Approach Combined with LES; 3. An Original Approach to Extract Momentum and Heat Transfers from Particle-Resolved Simulations of Particulate Flows; 4. Interfaces and Critical Fluids; 5. Shear Induced Anomalies in the Brownian Motion of Particles in Strongly Fluctuating Near-Critical Mixtures; and 6. Basics on Interfaces in Combustion.

In Volume 2, we examine cases that involve 1D, 2D or 3D *manifolds* in gaseous and liquid physical states, supercritical fluids, and single- or multi-phase systems that may be pure or mixed. This volume also consists of six chapters, which are as follows:

1 To consult these articles, see: https://www.openscience.fr/Thermodynamique-des-interfaces-et-mecanique-des-fluides.

Chapter 1 (J.-P. Caltagirone and S. Vincent) describes certain aspects of turbulence in discrete mechanics. The first part of this chapter is a brief description of the physical model associated with discrete primal and dual geometric topologies. The last section is devoted to simulating the turbulent channel flow with a turbulent Reynolds number of $Re_\tau = 590$.

Chapter 2 (R. Prud'homme) focuses on atomization in an acceleration field. The atomization of liquid jets injected into gaseous flows is an important problem that arises in combustion. It is especially important to know the size distribution of droplets to predict the evolution of the obtained sprays. In certain cases, an initial Kelvin–Helmholtz instability, created by vortices generating an acceleration field, becomes the source of a second instability known as the Rayleigh–Taylor instability, which determines the size of the droplets formed through this instability.

Chapter 3 (T. Chakkour) explores numerical studies of pipes with sudden contraction using *OpenFOAM*, and focuses on modeling that will be useful for engines and automobiles, bringing together models with 1D/3D couplings and revealing zones with strong gradients.

Chapters 4 (R. Prud'homme and K. Anani) and 5 (R. Prud'homme, K. Anani and M.N. Hounkonnou) study the evaporation of droplets subject to HF (high-frequency) perturbations, a possible cause of instabilities in injection engines. The Heidmann model, which replaces the droplets in motion in the combustion chamber by a single continuously fed droplet, is *complexified* by taking into account the finite conduction heat transfer phenomenon. This accentuates the damping of oscillations in the evaporation rate. We study the temperature field within the drop, as well as the response factor of a spherical drop to the high-frequency pressure field created by the engine. A pastille-shaped model of the drop is used, replacing the spherical drop, which makes it possible to go further with linearized calculations.

Finally, Chapter 6 (F. Raveloson, D. Tomboravo and R. Vony) presents the study of the interface, which is the surface of rotor blades in a Savonius wind turbine. A non-stationary and 3D flow is studied.

Roger PRUD'HOMME
Stéphane VINCENT
Christian CHAUVEAU
Mahouton Norbert HOUNKONNOU
November 2021

1

Turbulent Channel Flow to $Re_\tau = 590$ in Discrete Mechanics

Discrete mechanics based on an acceleration-potential formulation presents itself as an alternative to Navier–Stokes equations; the acceleration on an edge is the sum of a direct contribution due to the difference of the scalar potential at the ends and of another induced by the circulation of the vector potential on a dual contour. The law of dynamics becomes a law of conservation of acceleration on the segment considered; acceleration is written as the sum of two components, one divergence-free and another curl-free. This equation of motion makes it possible to apprehend turbulent flows in direct simulation, taking into account the propagation of longitudinal and transverse waves. The first part of this chapter is a brief description of the physical model, closely associated with discrete primal and dual geometric structures.

The last section is devoted to simulating the turbulent channel flow at a turbulent Reynolds number of $Re_\tau = 590$. It is demonstrated that discrete mechanics make it possible to accurately recover the mean velocity profiles of reference DNS and also to provide scale laws of the whole mean velocity profile from the wall to the center of the channel.

1.1. Introduction

Starting from the first drawings of Da Vinci (Lumley 1997), when one is interested in turbulence and its understanding, it can be chosen to adopt different complementary scientific approaches that have historically been experimental (Reynolds 1883), theoretical (von Kàrmàn 1948; Kolmogorov 1991) or numerical (Deardorff 1970). In the framework of this chapter, we choose the prism of the formulation of the models in fluid mechanics to

---

Chapter written by Jean-Paul CALTAGIRONE and Stéphane VINCENT.

propose a certain number of remarks and discussions related to turbulence and the phenomena it is sensitive to. If we rely here on the importance and meaning of models in the analysis and understanding of turbulence, this is particularly because new formulations of fluid mechanics equations have been published in recent years, more specifically those associated with discrete mechanics (DM) (Caltagirone 2019; Caltagirone and Vincent 2020).

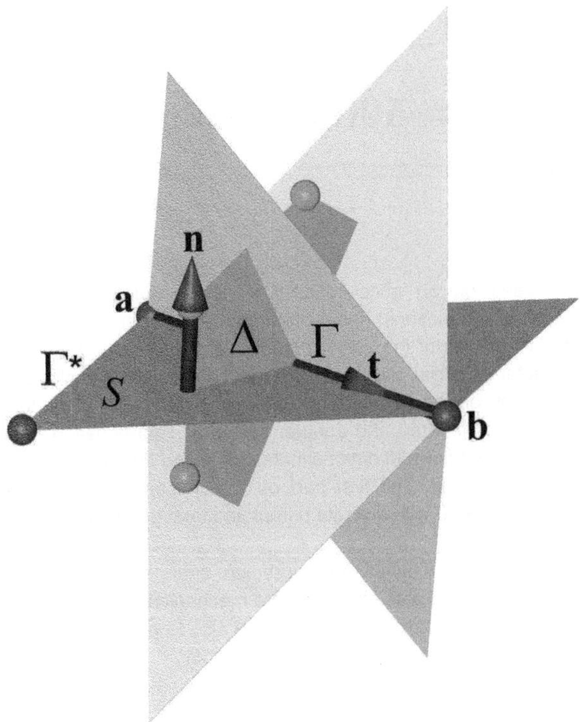

Figure 1.1. *Discrete geometric structure: a set of primitive planar facets S are associated with the segment Γ of unit vector $\mathbf{t}$ whose ends a and b are distant by a length d. Each facet is defined by a contour Γ^* oriented according to the normal $\mathbf{n}$ such that $\mathbf{n} \cdot \mathbf{t} = 0$; the dual surface Δ connecting the centroids of the cells is also flat. For a color version of this figure, see www.iste.co.uk/prudhomme/fluid2.zip*

From a general point of view, the notion of a discrete medium is directly derived from the principle of relativity of velocity and the weak equivalence principle: gravity accelerates all objects regardless of their mass or the materials from which they are made. All the contributions brought by the mechanical effects such as viscosity, compression or inertia, as well as all the source terms, are written as the sum of a free divergence term and a zero rotational term following the Helmholtz–Hodge decomposition. The motion equations of discrete mechanics reveal the role played by two quantities, namely the scalar and vector potentials, both associated with acceleration, a quantity considered as absolute. Even though the results obtained with this set of equations are generally the same as those of the continuous media provided by the Navier–Stokes equations, many formal differences exist. In particular, the density, as such, has disappeared from the momentum equation in favor of a scalar potential being the ratio between pressure and density. Many properties of the continuum are recovered intrinsically by the discrete mechanics formalism, in particular the conservation of mass, rotational or kinetic energy to cite a few. Trying to make the link between discrete or continuum mechanics formulations and analyzing or understanding of turbulence appears to us a relevant issue to be tackled.

Based on formulation differences, some features of discrete mechanics directly impact the understanding and modeling of turbulence. One of them relates to the value of the compressive viscosity for fluids, which is classically fixed by Stokes' law and is clearly questioned (Gad-el-Hak 1995; Rajagopal 2013). The treatment of the pressure in direct numerical simulation or the turbulent pressure in statistical turbulence modeling is directly related to the value of this compression viscosity, especially at the time scales of the small turbulent structures. Another aim of this chapter is to convince the reader that despite the fundamentally different nature of the discrete mechanics formulation, classical turbulent flow characteristics are nicely recovered and modeling issues can be revisited thanks to the DM approach.

This chapter is structured as follows. Section 1.2 provides a synthetic description of geometrical features of discrete mechanics together with a review of mathematical formulations, modeling issues and standard numerical aspects. The numerical simulation of the turbulent planar channel flow at a turbulent Reynolds number of $Re_\tau = 590$ (Moser *et al.* 1999; Denaro *et al.* 2011) is then investigated in section 1.5. In particular, comparisons to DNS and LES results of the literature are discussed and fitting laws for the mean velocity profile from the wall to the center of the channel are obtained.

1.2. Discrete mechanics formulation

The derivation of the equation of motion is carried out on a segment Γ, shown in Figure 1.1, from the law of discrete dynamics $\gamma = -\nabla \phi + \nabla \times \psi$, where γ is the proper acceleration of the material medium and the second member represents the imposed accelerations. Integration between the a and b ends of the Γ segment gives:

$$\int_{\Gamma} \gamma \cdot \mathbf{t} \, dl = - \int_{\Gamma} \nabla \phi \cdot \mathbf{t} \, dl + \int_{\Gamma} \nabla \times \psi \cdot \mathbf{t} \, dl \qquad [1.1]$$

The discrete motion equation is derived from the conservation equation of acceleration by expressing the deviations of potentials ϕ and ψ as a function of velocity $\mathbf{V}$. These "deviators" are obtained on the basis of the physical analysis of the storage-destocking processes of compression and shear energies; the first is written as the divergence of velocity, and the second as a dual curl of velocity. The physical modeling of these terms is developed in a book devoted to discrete mechanics (Caltagirone 2019) and a recent article on the subject (Caltagirone and Vincent 2020).

The vectorial equation of the movement and its upgrades are written as:

$$\begin{cases} \gamma = -\nabla \left(\phi^o - dt \, c_l^2 \, \nabla \cdot \mathbf{V} \right) + \nabla \times \left(\psi^o - dt \, c_t^2 \, \nabla \times \mathbf{V} \right) + \mathbf{g} \\ \alpha_l \, \phi^o - dt \, c_l^2 \, \nabla \cdot \mathbf{V} \longmapsto \phi^o \\ \alpha_t \, \psi^o - dt \, c_t^2 \, \nabla \times \mathbf{V} \longmapsto \psi^o \\ \mathbf{V}^o + \gamma \, dt \longmapsto \mathbf{V}^o \end{cases} \qquad [1.2]$$

The quantities ϕ^o and ψ^o are the equilibrium potentials; they represent the storage of compression and shearing energies between the instant of initial mechanical equilibrium and that at the instant t. They are written:

$$\phi^o = - \int_0^t c_l^2 \, \nabla \cdot \mathbf{V} \, d\tau; \qquad \psi^o = - \int_0^t c_t^2 \, \nabla \times \mathbf{V} \, d\tau \qquad [1.3]$$

The longitudinal and transverse celerities c_l and c_t are intrinsic quantities of the medium that can vary according to physical parameters. The terms

$dt\, c_l^2\, \nabla \cdot \mathbf{V}$ and $dt\, c_t^2\, \nabla \times \mathbf{V}$ are, respectively, the deviators of the compression and shear effects. The second member is thus composed of two oscillators ϕ^o and ψ^o, which represent energies per unit mass, exchanging these with their respective deviators. The two terms in gradient and in dual rotation are orthogonal and cannot exchange energy directly; if an imbalance due to an external event occurs for one of these effects, then the acceleration is changed and the energy is consequently redistributed to the other term. The acceleration **g** represents gravity or any other source quantity and will also be written in the form of a Helmholtz–Hodge decomposition.

The physical parameters α_l and α_t are the attenuation factors of the compression and shear waves, respectively. They also depend only on the medium considered; for example a Newtonian fluid retains the shear stresses only for very weak relaxation time constants, of order of magnitude of 10^{-12} s and the factor α_t can be taken as zero. The updating of potentials at time $t + dt$ is thus affected by these coefficients ranging between zero and unity. The velocity and possibly the displacement **U** are upgraded in turn. In the case where the density is not constant, it is also updated using the mass conservation in the form $\rho = \rho^o - dt\ \rho^o\ \nabla \cdot \mathbf{V}$; this quantity is only an explicit function of the divergence of velocity.

It should be noted that ϕ^o and ψ^o are energies per unit mass, each term reflecting the behavior of the medium or particle with respect to longitudinal and transverse waves. The term $\left(\phi^o - dt\, c_l^2\, \nabla \cdot \mathbf{V}\right)$ represents the longitudinal wave where ϕ^o accumulates the compression energy contained in the deviator term or restores it over time according to c_l. Similarly, $\left(\psi^o - dt\, c_t^2\, \nabla \times \mathbf{V}\right)$ is the oscillator corresponding to the velocity transverse waves c_t.

The time interval between two observations of the physical system in equilibrium is arbitrary, for example $dt = 10^{20}$ s to obtain a stationary state, whereas $dt = 10^{-20}$ s for the study in direct simulation of γ rays. The solution does not depend on this parameter if it is adapted to the mimicked physics; in the general case, the solution is of order 2 in space and time. Since all the terms are implicit on the variable velocity, or linearized as the terms of inertia, the system [1.2] is particularly robust.

The system [1.2] is generic and makes it possible to mimic the phenomena of solid mechanics, fluid mechanics, electromagnetism or optics, without any modification; the properties, $(c_l, c_t, \alpha_l, \alpha_t)$ are of course specific to the media considered and can vary over several tens of orders of magnitude. In fact it is, for example, the product $dt c_l^2$ that is significant of the phenomenon considered. Thus, for a gas flow, the time to be retained is $dt \approx 1/c_l^2 \approx 10^{-5}\, s$, whereas for visible light, it is necessary to keep a lapse of time small $dt \approx 1/c_l^2 \approx 10^{-17}\, s$. The resolution of this system can be achieved with the usual variables of each of the concerned domains or in terms of potentials. It is possible to return to the classical variables of mechanics at any time, fluid and solid, or electromagnetism, but the return to the variables of the represented domains is however not required, the solution of a problem depends in fact only on variables $(\mathbf{V}, \phi, \psi)$ and of course the intrinsic properties of studied media; we note that the latter often involved in the form of products of properties conventionally used. Similar to the velocity, the potentials ϕ and ψ must be advected; the writing of the system [1.2] makes it possible to separate the Lagrangian phase of upgrades for the potentials to the Eulerian advection phase, which has a completely different physical meaning.

Several research routes may be investigated in the future to highlight the properties of the equation of discrete motion in relation to turbulent flows: (i) the interaction of unsteady flows and longitudinal and also transverse waves; (ii) the specific effects of inertial terms, which also decompose according to two divergence-free and curl-free contributions; and (iii) modeling of large scales in a discrete context. Only the terms of diffusion of the discrete motion equation are put to the test in the case of a turbulent channel.

1.3. Turbulent flow in channel

1.3.1. *Analysis of a turbulent flow in a planar channel*

The majority of turbulence models are first validated against experimental results and then by direct simulations at moderate Reynolds numbers. More recently, databases have been compiled, allowing us to explore such flows in finer detail and develop more representative models, including problems where the production of turbulence arises from mechanisms other than just shearing.

We shall consider two databases in order to formulate the discrete model of a turbulent flow: the contributions of seven authors to a comparison by Denaro *et al.* (2011) with a turbulent Reynolds number of $Re_\tau = 590$, and earlier simulations by Moser *et al.* (1999) for the same problem.

The comparison by Denaro mainly focuses on the results of large-eddy simulations (LES); however, the model-free results are also presented, which are obtained for meshes with points in the viscous boundary sublayer as subgrid-scale direct simulations. This will be sufficient for an initial analysis of the discrete model. Details of the configuration are stated in the synthesis (Denaro *et al.* 2011).

Figure 1.2 shows the average velocity profiles in wall coordinates $u^+ = f(y^+)$. We can identify zones classically described in the literature *a priori*: the viscous sublayer for $y^+ < 5$, the buffer zone and the so-called logarithmic zone for $y^+ > 30$.

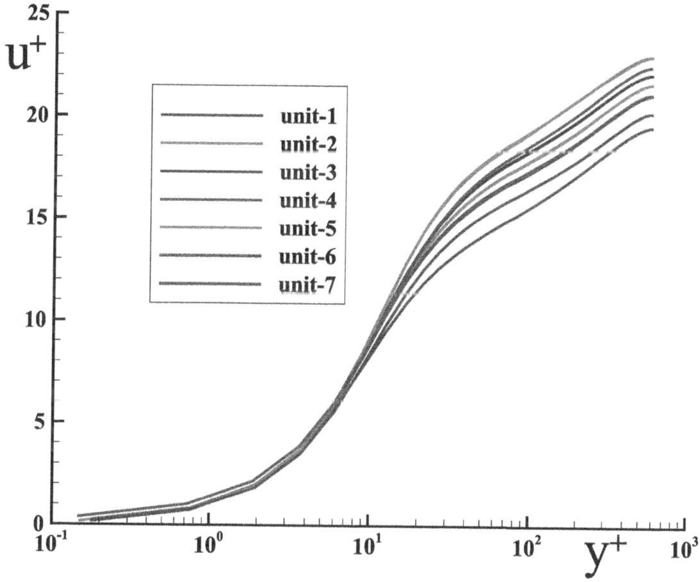

Figure 1.2. *Turbulent channel with $Re_\tau = 590$; average velocity profiles in reduced coordinates $u^+ = f(y^+)$, deduced from the various contributions to Denaro* et al. *(2011). For a color version of this figure, see www.iste.co.uk/prudhomme/fluid2.zip*

Starting from these average velocity profiles, we can characterize other zones in terms of the operators of discrete mechanics, including the dual curl of the primal curl $\mathbf{r} = -\nabla \times \nabla \times \mathbf{V}$. This vector points along the edge Γ with unit vector $\mathbf{t}$. The vector $\mathbf{r}$ is therefore collinear with the component $\mathbf{V} = \mathbf{W} \cdot \mathbf{t}$, where $\mathbf{W}$ is the velocity in a Galilean frame of reference. In practice, $\mathbf{V}$ can be viewed either as a vector or as a scalar defined on the edge Γ. This concept generalizes the ideas of staggered mesh introduced by Harlow and Welch (1965) for Cartesian meshes to arbitrary generalized polyhedral structures.

Figure 1.3a plots the variations of $\mathbf{r} = -\nabla \times \nabla \times \mathbf{V}$ as a function of the wall coordinates y^+. We can distinguish four zones separated by the values $y^+ = \delta_v$, $y^+ = \delta_l$, $y^+ = \delta_n$ as follows:

– $y^+ < \delta_v$, a layer where only the viscosity effects are present. The value of $\mathbf{r} = -\nabla \times \nabla \times \mathbf{V}$ is of order 1. It is difficult to truly characterize the behavior of this function from this example since the mesh is relatively coarse. But in general, the law of the velocity profile is known to be almost linear, $u^+ \approx y^+$;

– $\delta_v < y^+ < \delta_l$, a zone that can be calculated very precisely by simulations, where the viscosity effects remain dominant but the turbulence is no longer negligible. This zone is characterized by a minimum of the function $\mathbf{r} \cdot \mathbf{t}$ at $y^+ = \delta_r = 9$. While bearing in mind the relative inaccuracy in the definition of the zones δ_v, δ_l and δ_n, we can expect this value of $\delta_r = 9$ to hold more generally. This will need to be confirmed by analyzing other turbulent flows with different turbulent Reynolds numbers Re_τ;

– $\delta_l < y^+ < \delta_n$, a zone where the turbulence is not yet homogeneous, with some remaining viscosity effects, and where $\delta_l = 30$ is approximately equal to the inverse of the curvature of $\mathbf{r}$. On either side of this boundary, the curvature of $\mathbf{r}$ appears to be constant;

– $y^+ > \delta_n$, a region sometimes called the kernel where the turbulence is fully developed. Above $\delta_n = 200$, the value of $\mathbf{r} \cdot \mathbf{t}$ is constant. Since the curvature of the velocity profile is small in this region, direct simulations are less accurate and depend on the degree of convergence of the turbulent statistics.

The choice of values of δ_v, δ_l and δ_n will of course affect the solution, but contributions based on very different numerical methodologies (research codes, commercial software, etc.) nonetheless find very similar solutions.

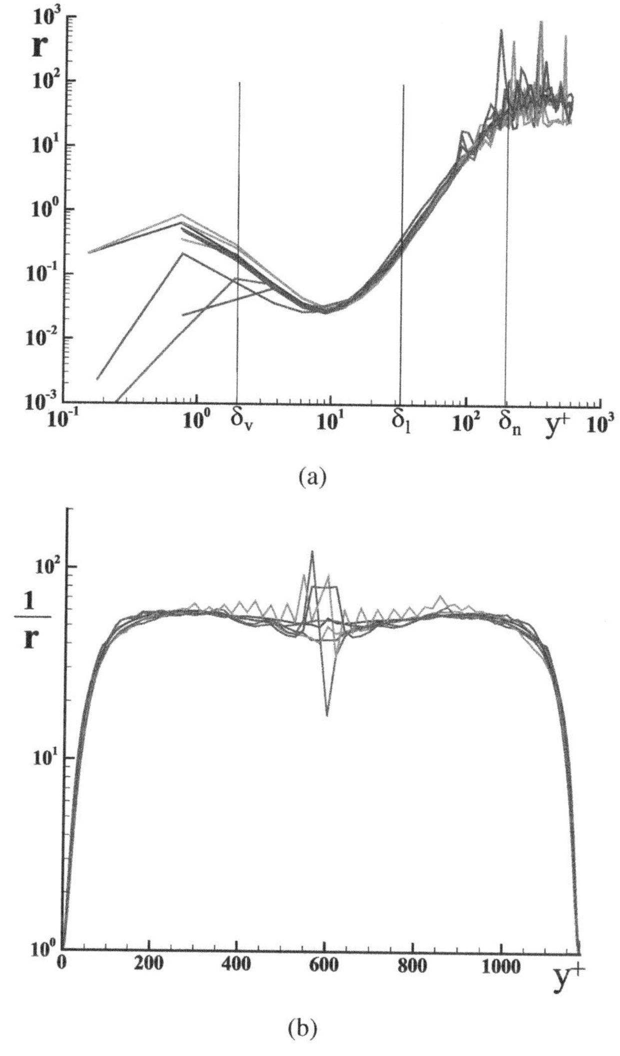

Figure 1.3. *(a) Turbulent channel with* $Re_\tau = 590$ *(Denaro* et al. *2011),* $\mathbf{r} = f(y^+) = -\nabla \times \nabla \times \mathbf{V}$ *in logarithmic coordinates. (b) The function* $1/\mathbf{r} = f(y^+) = 1/\nabla \times \nabla \times \mathbf{V}$ *is plotted across the width of the channel, reproduced from Denaro* et al. *(2011). For a color version of this figure, see www.iste.co.uk/prudhomme/fluid2.zip*

We shall choose the following boundaries between the regions described above: $\delta_v = 2$, $\delta_l = 35$ and $\delta_n = 200$. These values allow us to reproduce the average velocity profile of the contributor *unit*-2 very accurately. However, the variations of these boundaries are not as accurate at representing some of the other average profiles. At this stage, it would be difficult to fix the boundaries more precisely without having better convergence towards a unique solution. Once fixed, they can be viewed as universal constants.

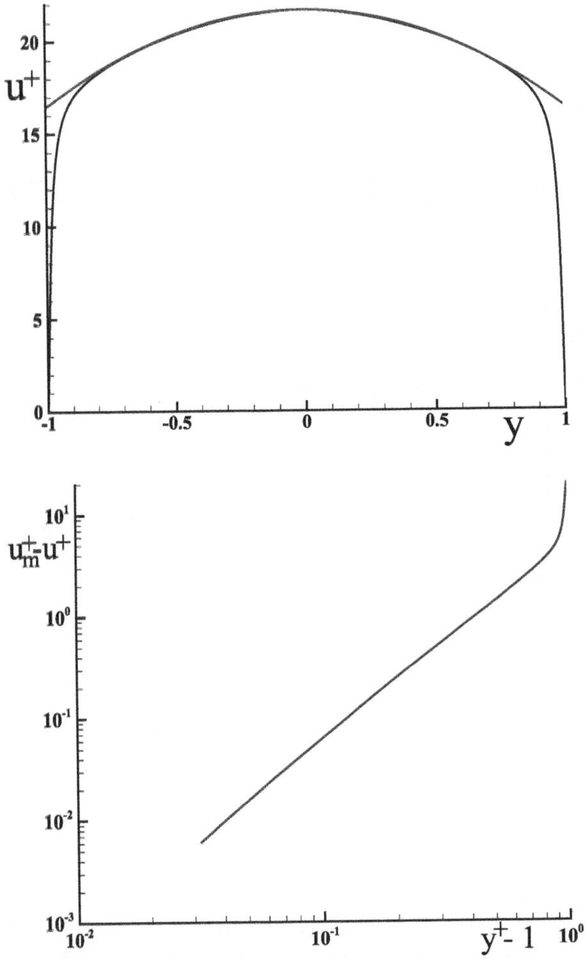

Figure 1.4. *Turbulent channel with $Re_\tau = 590$, results of unit-2 (Denaro et al. 2011). The average velocity profile $u^+ = f(y)$ is shown on the left, together with a parabolic profile for comparison. On the right, the same profile is plotted in logarithmic coordinates in the left half of the channel. For a color version of this figure, see www.iste.co.uk/prudhomme/fluid2.zip*

To gain a better understanding of the behavior of the solution in the turbulent kernel, we can plot the function $1/\mathbf{r} = 1/\nabla \times \nabla \times \mathbf{V}$ as a function of y^+ in linear coordinates. Figure 1.3(b) shows that this function is practically constant in the center of the channel. Of course, this is the region where the mesh is coarsest and the numerical errors are largest. It is also more difficult to ensure the convergence of the turbulent statistics in this region.

To analyze the behavior of the mean flow more finely, let us compare the velocity profile of the contribution *unit*-2 against a parabolic profile. A regression of the profile $u^+ = f(y)$ gives us the curvature κ, which is constant. Figure 1.4 shows a comparison with the parabolic profile. The graph of function $u^+ = f(y)$ on half of the channel shows that this function is strictly of order 2. The same procedure can be extended to each of the other contributions – only the curvature itself depends on the individual results. Table 1.1 lists the values of the curvature κ for each contribution.

Contribution	Curvature κ
unit-1	-5.35
unit-2	-5.21
unit-3	-5.17
unit-4	-5.25
unit-5	-4.89
unit-6	-5.31
unit-7	-5.24
MKM99	-6.6

Table 1.1. *Curvature in the zone $y^+ > \delta_n$ calculated by model-free simulations (Denaro* et al. *2011) and a direct simulation (Moser* et al. *1999)*

The table also shows the curvature found by the direct simulation by Moser *et al.* (1999). It is notably higher than the values of the model-free simulations from Denaro *et al.* (2011). At this stage, the most important conclusion is that the turbulent kernel can indeed be associated with an average velocity profile with constant curvature κ.

To determine more accurate universal values for the boundaries δ_v, δ_l and δ_n, we need an extremely accurate set of databases. Although the mean velocity profile itself seems very smooth, the values of the first- and second-order operators vary strongly, even in the direct simulations. The databases that are currently available are already sufficient to begin interpreting and analyzing the average profiles to establish a discrete-mechanical model of turbulence.

1.3.2. *Model of the turbulence in discrete mechanics*

One of the fundamental hypotheses of discrete mechanics is that the pressure effects are separable from the viscosity effects. Another key assumption is that the sum of all forces per unit mass can be expressed as two potentials, a scalar potential for the pressure stress and a vector potential for the shear-rotation stress $\gamma = -\nabla \phi^o + \nabla \times \psi^o$, where $p^o = \rho_v \, \phi^o$ is the mechanical equilibrium pressure and $\omega^o = \rho_v \, \psi^o$ is the equilibrium rotation stress. The equilibrium pressure and shear-rotation stresses are accumulators of the pressure and viscosity forces.

Recall the discrete equations established earlier:

$$\begin{cases} \dfrac{\partial \mathbf{V}}{\partial t} + \dfrac{1}{2} \nabla \left(\|\mathbf{V}\|^2 \right) - \dfrac{1}{2} \nabla \times \left(\|\mathbf{V}\|^2 \mathbf{n} \right) = -\nabla \left(\dfrac{p^o}{\rho_v} - dt \, c_l^2 \, \nabla \cdot \mathbf{V} \right) \\ -\nabla \times \left(\nu \, \nabla \times \mathbf{V} \right) + \mathbf{g} \\ \\ p = p^o - dt \, \rho_v \, c_l^2 \, \nabla \cdot \mathbf{V} \\ \\ \rho = \rho^o - dt \, \rho_v \, \nabla \cdot \mathbf{V} \end{cases} \quad [1.4]$$

where ρ_v is the density associated with the inertial or gravitational mass, defined as a constant on the edge Γ, whereas ρ is the density defined on the vertices of the primal mesh. The system [1.4] is autonomous and does not need to be accompanied by any state equations. Only the physical parameters themselves must be known. It does not matter how these parameters are determined: from tables, correlations, databases or by any other means. These quantities are local and instantaneous and can depend on the variables themselves. After the system has been solved, the quantities (p, ρ) are advected at the velocity of the fluid. This step is best accomplished by a Lagrangian transport calculation. The term in $\|\mathbf{V}\|^2$ from equation [1.4] can be incorporated into the gradient, corresponding to the dynamic pressure, and associated with the accumulation pressure p^o to give the Bernoulli pressure.

The formulation of the system [1.4] is the most compatible with the original discrete perspective. The underlying idea is that the evolution of the physical system is observed as it passes from one mechanical equilibrium state to another. Interpreting the system in terms of motion in a Galilean frame does not change the equilibria themselves, which are entirely governed by the Lagrangian equations of motion. Note that the system [1.4] represents every kind of flow, mechanical behavior and wave, including shock waves. Solving this system by a direct simulation (DNS) gives all of the physical information that is expected – not only the velocity or the displacement fields but also the instantaneous pressure and shear stresses. In particular, the system represents turbulent flows if the spatial and temporal scales are chosen accordingly.

The spatial scale characterizing a turbulent flow becomes smaller as the Reynolds number of the flow increases. Interactions between different scales of turbulence generate an apparent diffusion that could potentially be used to define a turbulent kinematic viscosity ν_{dm} in the context of an ad hoc model. Note that the fluid is assumed to be Newtonian and the turbulent dynamic shear viscosity μ_{dm} is not an intrinsic property but depends on the characteristics of the flow.

This idea of apparent viscosity can be incorporated into the discrete model. Unlike the turbulent viscosity μ_t of a statistical model, the apparent viscosity is not defined in terms of an equilibrium between production and dissipation. The production does not appear anywhere directly; for example, it cannot be found in the Navier–Stokes equations. The viscosity represented by the quantity ν_{dm} is not an intrinsic property of the fluid, but it is nonetheless a physical quantity. Knowledge of this quantity can be exploited to model the turbulence realistically. The objective of this section is to apply the concepts of discrete mechanics to establish a model of turbulent flows phrased exclusively in terms of the averages of specific quantities, such as the velocity or the pressure stress.

1.3.3. *Application to a turbulent flow in a channel with $Re_\tau = 590$*

Historically, one of the most emblematic problems ever studied since the original experiments by Reynolds is the flow within a channel, which becomes turbulent at Reynolds numbers of the order of $Re \approx 2300$. Many experimental studies, direct numerical simulations (DNS), large-eddy simulations (LES) and even Reynolds-averaged Navier–Stokes equations (RANSE) statistical models have examined this problem, resulting in extremely accurate databases.

The problem setup is straightforward. A flow with a constant Reynolds number is considered within a channel with rectangular cross-section, driven by a fixed pressure difference δp in the x-direction. Since the length of the channel is assumed to be large relative to its hydraulic diameter, the average pressure gradient ∇p is constant along the channel. To simulate this problem, periodic boundary conditions are typically assumed in the x- and z-directions.

Each zone of the flow in the channel is modeled directly from equation [1.4]. The conclusions of our model will be fully consistent with the original dimensionless analysis that was performed to describe the behavior of this flow. But for now, these equations are already fully representative of any laminar or turbulent flow; the analysis that we shall perform below is based on the equilibrium between the pressure and the curl of the velocity brought by the discrete mechanics equations. In particular, between any two layers of the fluid [1.1] and [1.2], the following equality holds:

$$-\nu_1 \left(\nabla \times \mathbf{V}\right)_1 = -\nu_2 \left(\nabla \times \mathbf{V}\right)_2. \qquad [1.5]$$

This condition is intrinsically satisfied. It replaces the equality of shear stresses described by the Cauchy tensor. It also holds for two immiscible fluids in problems with multiphase flows.

Since we are only interested in finding the steady average flow, the partial derivative of the velocity with respect to time is zero. The two inertial terms are also zero. This property is of course not satisfied in direct simulations. The equations of motion characterizing established turbulent motion in a channel with constant cross-section may therefore be expressed as follows in terms of average values:

$$-\nabla p^o - \nabla \times \left(\mu_{dm} \nabla \times \mathbf{V}\right) = 0. \qquad [1.6]$$

The Helmholtz–Hodge theorem defines the gradient of a scalar function and the curl of a vector quantity as orthogonal. The two fields ∇p and $\nabla \times \boldsymbol{\omega}$ are equal and orthogonal, and so must both necessarily be equal to a constant. Since the average pressure field is only a function of x and the shear field is only a function of y, equation [1.6] could, in principle, be reduced to partial derivatives along x and y. However, we shall avoid this reduction, since these operators have a physical meaning that disappears if we replace $\nabla \times \mathbf{V}$ by $d\mathbf{V}/dy$, which is just part of the component of the curl.

This zone of the flow is defined by the reduced coordinates $y^+ = y\, u_\tau/\nu < \delta_v$, where $u_\tau = \sqrt{\tau_w/\rho}$ is the shear velocity and ν is the kinematic viscosity. The reduced velocity is classically defined by $u^+ = u/u_\tau$. The characteristic length δ_v of this zone can thus be determined by studying the turbulent boundary layer of a planar sheet, where it is typically equal to $y^+ \approx 7$.

Within this layer, the viscous effects are dominant and any remaining velocity is maintained by the outer layers. This is a Couette flow generated by the adjacent fluid, known as the buffer zone. However, in a planar Couette flow, the curl is constant and the pressure gradient is zero. This is not the case for the average gradient ∇p^o. The only way to recover Couette flow in this zone is to take $\nabla \times \boldsymbol{\omega}^o = \nabla p^o$ and $\mu_{dm} = \mu$; equation [1.6] then becomes:

$$\nabla \times \mathbf{V} = 0, \qquad [1.7]$$

i.e. $u^+(y^+) - \mathbf{V} \cdot \mathbf{e}_x = y^+$. This recovers the classical linear law, except that the slope is imposed by the external flow.

From a physical point of view, the imposed pressure in any given cross-section is determined solely by the constant gradient ∇p^o. This is especially true if the thickness δ_l is very low. Without the term in $\nabla \times \mathbf{V}$, the gradient in the viscous sublayer cannot be eliminated, and the law $u^+ = f(y^+)+$ is not linear but quadratic. Hence, the shear stress is steady and accumulates in $\boldsymbol{\omega}^o$ according to the law $\boldsymbol{\omega}^o(y^+) = y^+\, \mathbf{e}_z$.

In boundary layer theories, these two regions – the buffer zone and the logarithmic zone – are separate. This is not the case here, in discrete mechanics. The buffer zone is not only a coupling zone but also plays an essential role in the equilibrium between the pressure and the viscous stresses. These two effects are of the same order and modeling them can be particularly tricky. We shall assume that the flow is unsteady in each zone, allowing the accumulated viscous stresses to be destroyed, with $\alpha_t = 0$. Consequently, the flow is generated by the constant pressure gradient $\nabla p^o = \delta$. The equations of motion now become:

$$-\nabla p^o - \nabla \times (\mu_{dm}\, \nabla \times \mathbf{V}) = 0, \qquad [1.8]$$

which gives:

$$\frac{d}{dy^+}\left(\mu_{dm}\,\frac{du^+}{dy^+}\right) = \delta. \qquad [1.9]$$

Since δ is a constant, we must have $\mu_{dm} = y^{+2}$, since the product $y^+ \cdot du^+/dy^+$ is constant, as established by dimensionless analysis and corroborated by direct simulations.

The solution of the above equations with $\mu_{dm} = y^{+2}$ is:

$$u^+ = \delta\, Ln(y^+) - \frac{a}{y^+} + b. \qquad [1.10]$$

The logarithmic law is recovered as expected, but there is also an additional term in $1/y^+$ that increases in significance as y^+ approaches δ_v and becomes negligible as y^+ approaches $\delta_l \approx 140$ at the end of the logarithmic zone.

The constants a and b can be deduced from the profiles calculated by DNS, but they are no longer relevant anyway after equation [1.8] is integrated directly. The coupling between the different zones is ensured by the implicit condition [1.5]. Hence, the apparent viscosity μ_{dm} is equal to μ in the viscous sublayer, increases slowly in the buffer zone and increases more rapidly in the logarithmic zone. These variations are fully determined by the apparent viscosity in these two zones, $\mu_{dm} = y^{+2}$.

In the kernel of the flow, after $y^+ = \delta_l \approx 140$, the turbulence is fully developed and the diffusion of momentum is maximal; in other words, μ_{dm} attains its highest values in this zone. In this region, the flow is mainly characterized by the pressure gradient. To satisfy equation [1.8], we must invoke the results of direct simulations, as well as the physical fact that this zone has a high mixing rate. The average velocity profiles can be very precisely represented by a second-order law that takes into account the symmetry associated with the axis of the channel. This law can be expressed in the form $u^+ = u^+_m - y^{+2}/\kappa$, where κ is the absolute value of the curvature of the velocity profile. This is a constant that depends on the turbulent Reynolds number Re_τ. This gives the value $\mu_{dm} = -\delta/2\,\kappa = \mu_m$ of the viscosity. The apparent viscosity is constant for a given turbulent Reynolds number.

Let us briefly present the results obtained by the discrete approach (Caltagirone 2019) for the flow in a channel with $Re_\tau = 590$. Consider a 3D channel with rectangular cross-section whose lateral dimensions in the x- and z-directions are larger than the vertical dimension in the y-direction. No-slip conditions are imposed on the horizontal walls; periodic boundary conditions are applied in the x-direction.

A constant pressure gradient $\delta p \, \mathbf{e}_x$ is maintained throughout the volume to fix the value of the average velocity $V_0 = \mathbf{V}_0 \cdot \mathbf{e}_x$. The dimensions and the velocity V_0 are chosen in such a way that the turbulent Reynolds number Re_τ is 590. Dimensionless quantities y^+ and u^+ are then derived from the vertical size of the channel and the local velocity based on V_0.

This configuration has been studied extensively since the early 1980s by performing direct simulations or large-eddy simulations. The results are broadly consistent at low Reynolds numbers like these. We can mention the work by Kim *et al.* (1987) or Moser *et al.* (1999), for example. More recently, other authors have published refined comparisons that were reviewed in a test case organized by Denaro in 2011. The results were updated in 2014 and are available in the literature (Denaro *et al.* 2011).

Our objective here is to develop a steady turbulence model for the averaged parameters that is based fully on the discrete-mechanical approach and which may be applied to the discrete equations arising from this model. Even though the Navier–Stokes equations lead to identical results when the physical properties are constant, they differ significantly from the discrete equations, meaning that a new model of turbulence is needed for the discrete approach. The concept of rotation effectively lies at the heart of the discrete theory, which describes viscous effects solely in terms of the vector potential $\boldsymbol{\omega}$ rather than a stress tensor.

Many of the approaches currently used to model the turbulence appear to conflate the diffusion of momentum and the viscous dissipation. These are two different effects, and most models from statistical modeling and LES are in fact based on the dissipation.

In discrete mechanics, the transfer of momentum within the fluid is governed by the dual curl and the shear-rotation stress $\boldsymbol{\omega}$ through the term $\nabla \times (\boldsymbol{\omega}^o - \mu \nabla \times \mathbf{V})$. The quantity $\boldsymbol{\omega}^o$ represents the storage of stresses of diffusion associated with μ, the shear-rotation stress, which is the asymptotic value of the instantaneous effects. For solids or fluids with very small time constants, this term can be written as $dt\ \mu_e$, where μ_e is the elastic shear modulus. In fact, this time constant corresponds to the period dt between two observations of the physical system.

In turbulence, this characteristic timescale becomes smaller as the Reynolds number becomes larger. The shear stresses are diffused by vortices, which creates an apparent diffusion phenomenon as a result of mixing. We can define viscosity arising from this turbulence but it represents a completely separate concept from the viscous dissipation, which does not appear directly in the equations of motion. It is neither a turbulent viscosity nor a subgrid-scale viscosity. We shall call this viscosity μ_{dm}, and it will, of course, depend on the local conditions of the flow.

The simulation was performed in a 2D domain with periodic conditions in x and a mesh with 100 cells in the x-direction and 64 cells in the y-direction. The initial conditions on $(p, \mathbf{V})$ were set to zero, and the pressure gradient was chosen in such a way that $Re_\tau = 590$. The steady solution was calculated up to the machine error. A one-dimensional model was also developed, and the results are identical. The steady version of the problem that is periodic in x effectively reduces to a one-dimensional equation along y.

The problem is modeled implicitly by layers using the term $\nabla \times (\boldsymbol{\omega}^o - \mu_{dm}\ \nabla \times \mathbf{V})$, which already appears in the equations of motion, based on the various zones of the flow:

- the laminar boundary sublayer for $y^+ < \delta_1$;
- the buffer zone for $\delta_1 < y^+ < \delta_2$;
- the logarithmic zone for $\delta_2 < y^+ < \delta_3$;
- the fully turbulent zone for $y^+ > \delta_3$.

The values of the boundaries δ_1, δ_2 and δ_3 are defined from values in the literature that are consistent with previously conducted direct simulations. A strictly linear law $u^+ = f(y^+)$ can only be obtained for the accumulation stress $\boldsymbol{\omega}^o$, since the constant gradient maintained within the laminar boundary sublayer is not compatible with a linear law. However, the thickness δ_1 is very small and the pressure is the same as in the buffer zone. In practice, the term $\boldsymbol{\omega}^o$ balances out the effect of the pressure gradient and the velocity law is perfectly linear in the boundary sublayer.

The physical explanation for this is as follows: within the laminar boundary sublayer, the flow is steady and μ has attained its asymptotic limit, known as the molecular viscosity. After this point, unsteady exchanges between vortices cause a small part of the rotation stresses to accumulate, but anything that accumulates is then immediately diffused by these same exchanges. This leads to larger viscosity values, $\mu_{dm} > \mu$.

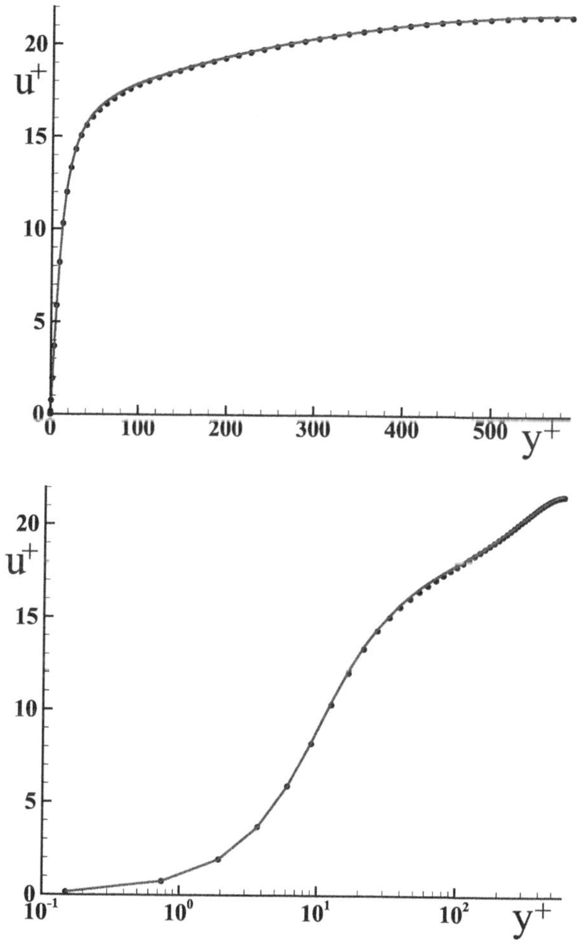

Figure 1.5. *Turbulent channel with $Re_\tau = 590$, comparison between the results of DNS (unit-2) (Denaro* et al. *2011) (solid line) and the discrete approach (Caltagirone 2019) (dots). The profile $u^+ = f(y^+)$ of the average velocity is shown in reduced coordinates on the left and logarithmic coordinates on the right. For a color version of this figure, see www.iste.co.uk/prudhomme/fluid2.zip*

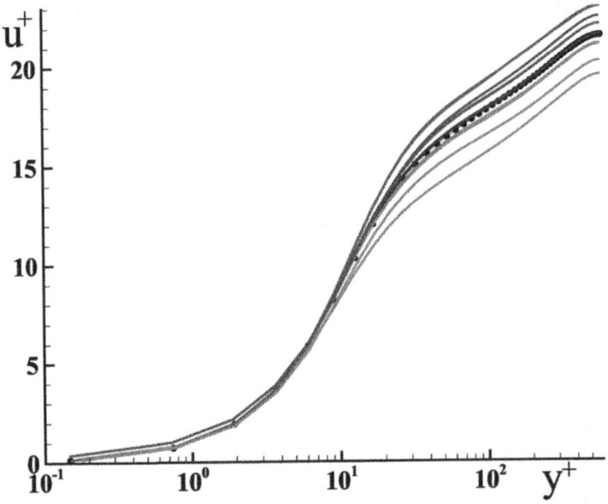

Figure 1.6. *Turbulent channel with $Re_\tau = 590$. Results of the TDM model (dots) compared to direct simulations reproduced from Denaro* et al. *(2011) (solid lines). For a color version of this figure, see www.iste.co.uk/prudhomme/fluid2.zip*

Figure 1.5 shows the average velocity $u^+ = f(y^+)$ in physical coordinates and in logarithmic coordinates. The distinct zones of the flow are clearly distinguishable. The results are compared against the results found by Germano, reproduced from Denaro *et al.* (2011). The objective of this section is to use physical concepts to establish a robust and reliable averaged model with as few parameters as possible that can be determined from direct simulations. Figure 1.6 plots the values of $u^+(y^+)$ over time obtained by the discrete model against the results of direct simulations recorded in Denaro *et al.* (2011). The points issued to the model accurately reproduce the evolution in dimensionless coordinates. The dispersion of the results of the direct simulations can be attributed to methodological differences in the authors' approaches, as well as refinements of the mesh near the walls.

Figure 1.7 gathers together the results for the reduced velocity μ_{dm}/μ as a function of the reduced position y^+ in physical coordinates and logarithmic coordinates. In the central zone, the ratio is essentially constant. The same is true in the viscous boundary sublayer, where the ratio is equal to 1. Between these two zones, the evolution is quadratic.

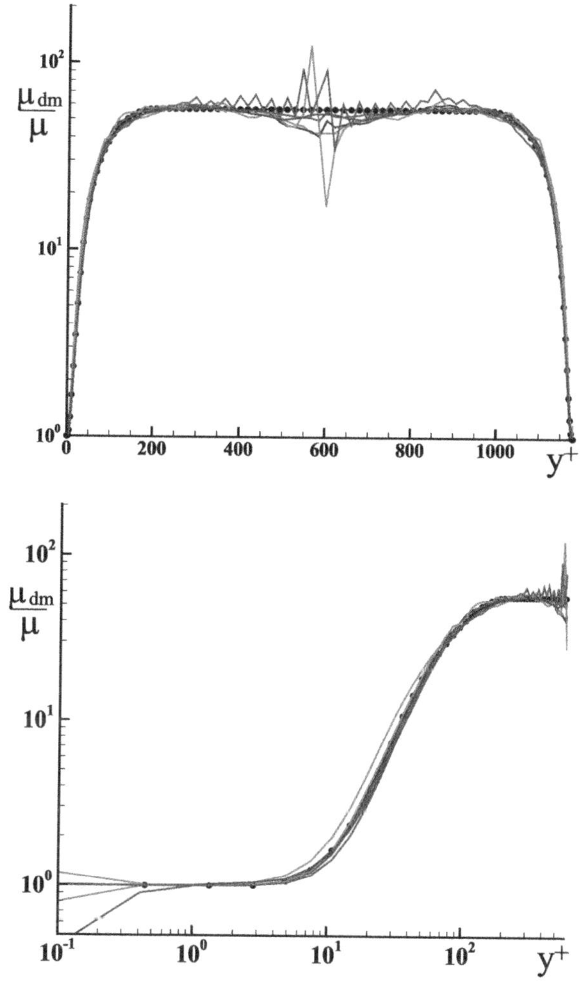

Figure 1.7. *Turbulent channel with $Re_\tau = 590$. The reduced turbulent viscosity μ_{dm}/μ is shown for the TDM model (dots) and the DNS by Denaro* et al. *(2011) (solid lines). For a color version of this figure, see www.iste.co.uk/prudhomme/fluid2.zip*

The simulations by Moser *et al.* (1999) are one of the possible references for this turbulent flow problem in the literature. Figure 1.8 plots the reduced velocity as a function of the reduced position in wall units, showing that these reference results are highly consistent with the results of the discrete model.

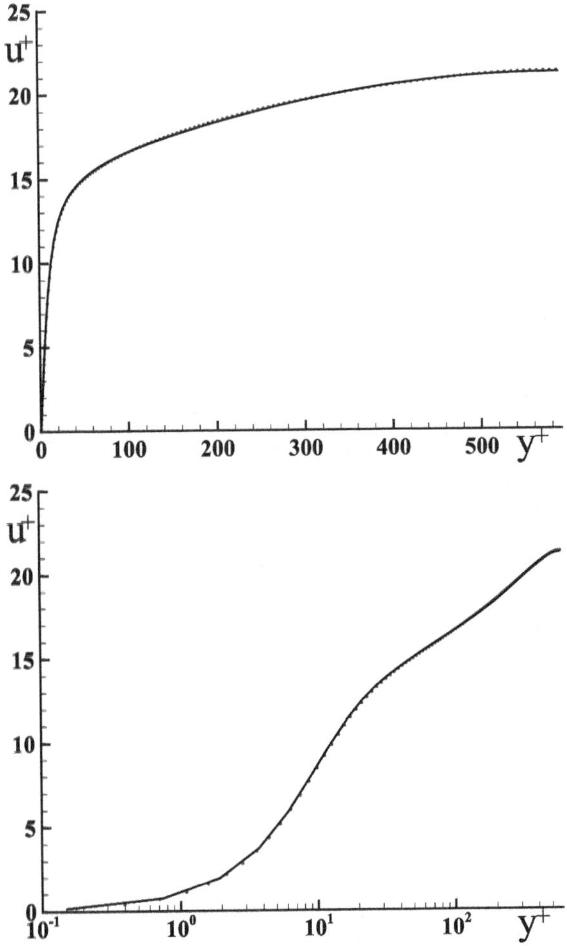

Figure 1.8. *Turbulent channel with* $Re_\tau = 590$, *results from MKM99 (Moser* et al. *1999) (solid line) and from the TDM model (dots). The average velocity profile* $u^+ = f(y^+)$ *is shown in wall units on the left and in logarithmic coordinates on the right. For a color version of this figure, see www.iste.co.uk/prudhomme/fluid2.zip*

We do not impose or even consider any laws on the velocity profiles, since they can be implicitly reproduced by choosing the viscosity μ_{dm} accordingly. So, which parameters do we need to model the flow in a turbulent channel, and hence to model an arbitrary problem? In the viscous boundary sublayer, the viscosity is given by μ, the molecular viscosity, but the thickness δ_v of this layer needs to be known. The intermediate layer requires us to know another thickness to determine the end of the logarithmic zone, and the turbulence zone presented above requires us to know κ, the curvature of the velocity profile in this zone. Therefore, at the very least, we need the three parameters δ_v, δ_l and δ_n, as well as κ itself. This model also needs to satisfy various constraints. In particular, it must naturally reduce to Poiseuille's solution for a laminar flow in the degenerate case. In this case, the curvature κ is given by $\kappa = -1/2\, \mu$, and the viscous sublayer and the intermediate zone vanish. The laminar flow occupies the whole channel.

Locally, the viscosity μ_{dm} can be calculated from the equilibrium between the pressure effects and the shear effects as follows in discrete mechanics:

$$\mu_{dm} = \frac{\nabla p}{\nabla \times \nabla \times \mathbf{V}}, \qquad [1.11]$$

where the denominator is implicitly understood as $\nabla_d \times \nabla_p \times \mathbf{V}$. The term $\nabla_d \times$ is the dual curl, and $\nabla_p \times$ is the primal curl. This is in fact just the curvature of the velocity profile.

The two vector quantities in equation [1.11] have the same direction as the unit vector $\mathbf{t}$ of the edge Γ. It therefore makes sense to define the scalar μ_{dm} on this same edge. Statistical models produce much more fragmented results involving the introduction of so-called universal constants. Furthermore, models based on Boussinesq's hypothesis and the concept of turbulent viscosity μ_t typically only hold in specific conditions – high Reynolds numbers, low Reynolds numbers, etc.

$$\begin{cases} y^+ < \delta_v, & \dfrac{\mu_{dm}}{\mu} = 1 & u^+ = y^+ \\[2ex] \delta_v < y^+ < \delta_l, & \dfrac{\mu_{dm}}{\mu} = 1 + 5\,(y^+ - \delta_v)^2 & u^+ = \delta\, Ln(y^+) - \dfrac{a}{y^+} + b \\[2ex] \delta_l < y^+ < \delta_n, & \dfrac{\mu_{dm}}{\mu} = \mu_m - (y^+ - \delta_n)^2 & u^+ = \delta\, Ln(y^+) - \dfrac{a}{y^+} + b \\[2ex] y^+ > \delta_n, & \dfrac{\mu_{dm}}{\mu} = \mu_m = -\dfrac{2\,\delta}{\kappa} & u^+ = u_m^+ - \dfrac{y^{+2}}{\kappa} \end{cases} \qquad [1.12]$$

The system [1.12] summarizes the law of the wall in each zone as a function of the reduced position. These zones are defined by the distances $\delta_v = 2$, $\delta_l = 35$, $\delta_n = 200$ and $\mu_m = -2\,\delta/\kappa$.

1.4. Conclusion

The brief description of the physical model based on a discrete approach to motion shows the essential differences between the Navier–Stokes equations and the equation of discrete mechanics. The latter makes it possible to replace the operators of classical mechanics, where the velocity gradient plays an important role in terms of both inertia and viscous diffusion, by simple operators of differential geometry. All physical variables and parameters are expressed with only two fundamental units, those of length and time. This alternative to the equations of classical mechanics is suitable for studying the propagation of longitudinal and transverse waves in turbulent flows. This concerns both small time scales and large time constants, in LES or for the statistical modeling of turbulence.

In the last section, the simulation of the turbulent channel flow at $Re_\mathcal{T} = 590$ demonstrated that the formulation of acceleration equations brought by discrete mechanics, which is intrinsically different from the momentum equations in the Navier–Stokes model, makes it possible to accurately recover the DNS and LES results of the literature (Moser *et al.* 1999; Denaro *et al.* 2011). Clear evidence of the physical meaning of DM for turbulence flow modeling was given. In addition, based on DM, the fitting of simulation results was possible for the mean streamwise velocity without any requirement for dimensional analysis considerations. The obtained law is valid from the wall to the center of the channel with a nice agreement with DNS reference values.

Future works and extension of DM use for turbulent motions are oriented in two directions: the analysis of well-known turbulent flows such as homogeneous isotropic turbulence (HIT), two-phase flows and jets, on the one hand, and the building of a new LES or RANS model based on the new DM formulation of motions, on the other hand.

1.5. References

Caltagirone, J.-P. (2019). *Discrete Mechanics: Concepts and Applications*. ISTE Ltd, London and John Wiley & Sons, New York.

Caltagirone, J.-P. and Vincent, S. (2020). On primitive formulation in fluid mechanics and fluid-structure interaction with constant piecewise properties in velocity-potentials of acceleration. *Acta Mechanica*, 231(6), 2155–2171.

Deardorff, J. (1970). A numerical study of three-dimensional turbulent channel flow at large Reynolds numbers. *Journal of Fluid Mechanics*, 41, 453–480.

Denaro, F.M., Abbà, A., Germano, M., Icardi, M., Marchisio, D., Rolfo, S., Lampitella, P., Colombo, E., Inzoli, F., Aprovitola, A., Marra, F.S., Ioveno, M., and Tordella, D. (2011). A comparative test for assessing the performances of large-eddy simulation codes. *Proceedings of XX AIMETA Conference*, Bologna, Italy, 12–15 September.

Gad-el-Hak, M. (1995). Questions in Fluid Mechanics: Stokes' hypothesis for a Newtonian, isotropic fluid. *Journal of Fluids Engineering*, 117(1), 3–5.

Harlow, F.H. and Welch, J.E. (1965). Numerical calculation of time-dependent viscous incompressible flow of fluid with a free surface. *Physics of Fluids*, 8, 2182–2189.

von Kàrmàn, T. (1948). Progress in the statistical theory of turbulence. *Proceedings of the National Academy of Sciences of the United States of America*, 34(11), 530–539.

Kim, J., Moin, P., and Moser, R.D. (1987). Turbulence statistics in fully developed channel flow at low Reynolds number. *Journal of Fluid Mechanics*, 177, 157–159.

Kolmogorov, A.N. (1991). The local structure of turbulence in incompressible viscous fluid for very large Reynolds numbers. *Proceedings of the Royal Society of London A*, 434, 9–13.

Lumley, J.L. (1997). Some comments on turbulence. *Physics of Fluids A: Fluid Dynamics*, 4(2), 203–211.

Moser, R.D., Kim, J., and Mansour, N.N. (1999). Direct numerical simulation of turbulent channel flow up to $Re_\tau = 590$. *Physics of Fluids*, 11(4), 943–945.

Rajagopal, K.R. (2013). A new development and interpretation of the Navier–Stokes fluid which reveals why the "Stokes assumption" is inapt. *International Journal of Non-linear Mechanics*, 50, 141–151.

Reynolds, O. (1883). An experimental investigation of the circumstances which determine whether the motion of water shall be direct or sinuous, and of the law of resistance in parallel channels. *Proceedings of the Royal Society of London*, 35(224–226), 84–99.

2

Atomization in an Acceleration Field[1]

Acceleration fields other than gravitational fields are sources of instabilities in liquid layers and liquid jets. We briefly examine two cases that lead to atomization at a liquid filament stage: ultrasonic vibrations, and parallel or concentric injector jets. The size distribution of drops thus formed is important in designing engines with efficient combustion. The influence of various parameters, including the density ratios of injector jets, is current the topic of research.

List of symbols

Symbol	Meaning	Symbol	Meaning
a	acceleration at the crest of a wave	Re	Reynolds number
$a_m(t)$	coefficient	S_m	eigenfunction of the Laplacian of the surface
b	thickness of the crest	t	time
C_{axi}	proportionality constant	T	reduced time
C_D	drag coefficient	U_i	velocity of the fluid i

Chapter written by Roger PRUD'HOMME.

1 This article is an updated version of an oral presentation by the author within a research group on Fundamental and Applied Microgravity, titled "Atomization in an Acceleration Field". Colloquium by the GDR MFA – Belgodère – October 11–14, 2016.

Symbol	Meaning	Symbol	Meaning
D, d	diameter of a drop	U_c	convection velocity of the wave
$D_{p,q}$	mean diameter associated with indices p, q	**V**	velocity vector
$D_{3,2}$	Sauter mean diameter	We	Weber number
e	injector output separation thickness	x, y	coordinates
E_*	energy involved in a firework	α	ratio b/λ_{axi}
E_*^c	critical energy	β	angle of the separation plate
f'	maximum acceleration of an oscillation	δ_G	vorticity thickness of the gaseous jet
F	force vector acting per unit volume	ϕ	velocity potential
g	acceleration due to gravity	λ_{axi}	Kelvin–Helmholtz wavelength
h	thickness of the liquid	λ_{RT}	Rayleigh–Taylor wavelength
		μ	coefficient of shear viscosity
H_G, H_L	thickness of gas and liquid jets, respectively, at the injector outlet	$\mu(p_m, q_m)$	coefficient of the Mathieu–Hill equation
i	fluid (liquid: i=L, gas: i=G)	ν	kinematic viscosity $\nu = \mu/\rho$
k	wave number	ρ	density
$k*$	eigenvalue for the Laplacian of the surface	σ	surface tension
M	ratio of the kinetic energies of gas/liquid	ω	pulsation
P	pressure	$\zeta(x, y, t)$	elevation of a disturbed surface in 2D configuration
p_m, q_m	frequency, amplitude in the Mathieu-Hill equation		

2.1. Introduction

2.1.1. *Two classic instabilities*

Bodies in motion follow Newton's law, giving the local equality between the applied force and the acceleration resulting from this force. The displacement of a liquid depends on the force fields applied to it through local contact (e.g. through friction), or at a distance (e.g. gravity, an electric field, a magnetic field or drag acceleration). A liquid structure at equilibrium or in regular motion can be destabilized under the influence of such fields.

The Rayleigh–Taylor instability, between two superimposed, horizontal liquid layers initially at rest, is well known. If the upper liquid layer is the denser layer, digitations of this liquid may spontaneously form within the liquid that is less dense (Figure 2.1a). The linearized analysis of this phenomenon develops a growth curve of small disturbances and makes it possible to determine the domain of instability, with or without the presence of the surface tension σ in the plane (k,ω_i) with the wave number depending on the frequency (Lamb 1945).

If these two superimposed liquids move with different velocities (U_1,U_2), we have Kelvin–Helmholtz instability (Chandrasekhar 1961) characterized by the fluid coiling along the surface, propagating itself with the convection velocity U_c. If the velocities are high and $U_1 > U_2$, we have:

$$U_c = U_1\left(1 - U_2/U_1\left(\sqrt{\rho_2/\rho_1}\right)\right)\left(1+\sqrt{\rho_2/\rho_1}\right)^{-1} \qquad [2.1]$$

where (ρ_1,ρ_2) are the respective densities of the fluids.

Figure 2.1b shows such coiling in the case of liquid oxygen that is injected into gaseous hydrogen where combustion occurs.

a)

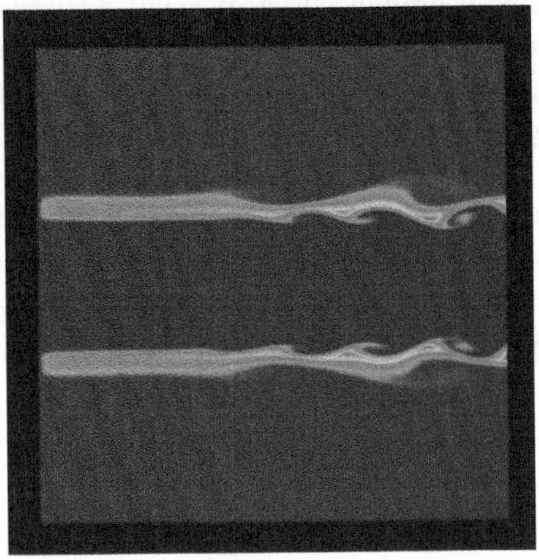

b)

Figure 2.1. *(a) Rayleigh–Taylor instability produced by carefully injecting hot, colored water into a glass of cold water. This configuration is initially stable, but as the colored water cools down and becomes denser than the water below it, it becomes unstable. The descending "mushrooms" are characteristic of the Rayleigh–Taylor instability, according to Jens Niemeyer (Niemeyer 2015). (b) Double Kelvin–Helmholtz downstream of a running* O_2*–*H_2 *injector (Nicoli* et al. *2003) (taken from the video kindly provided by C. Nicoli and P. Haldenwang). For a color version of this figure, see www.iste.co.uk/prudhomme/fluid2.zip*

2.1.2. *Atomization*

The atomization of a liquid is the transformation of a compact quantity of this liquid into droplets. This can occur in various forms and the process generally consists of several steps (Prud'homme 2014).

Azzopardi (1988) studied the formation of droplets in a pipe. Some of these droplets formed by a tearing away at the level of the liquid film present on the internal wall. The process is complicated by thermal exchanges associated with evaporation. The drops that are formed through various processes may transform into liquid pockets and burst, dispersing fine droplets – and there may be an intermediary ligamentous step.

Marmottant and Villermaux (2004) experimentally obtained ligaments (or fibers) by more or less quickly pulling out a vertical tube that was initially plunged into a liquid. The studies they carried out demonstrated that the internal structure, made up of small balls of stretched liquid ligaments, is governed by statistical laws just like molecular groups in statistical thermodynamics.

Antkowiak *et al.* (2007) experimentally studied the formation of a filament obtained through the shock of a test tube against a solid surface. Among other things, they demonstrated the role of the curvature of the liquid–gas interface in obtaining the ligament.

More recently, Inoue *et al.* (2017) studied the difficult problem of cascade fragmentation in fireworks. In this context, they focused on the evolution of a droplet disturbed by the bursting of an internal gaseous cavity. They thus showed that the drop may oscillate around its initial spheroidal form when the energy involved, E_*, is smaller than a certain critical value, E_*^c. When E_* is sufficiently high, the drop may then deform into a ligament that becomes elongated enough to break up.

In the following sections, we will examine situations that lead to the formation of drops with the following characteristics:

1) an acceleration field with a mechanical origin, in the presence or absence of gravity;

2) processes that involve the formation of ligaments (fibers or filaments).

Atomization is a result of developing instabilities. We give two examples of droplets being generated as follows:

– parametric instability obtained through vibrations that are normal to the liquid layer;

– a Rayleigh–Taylor instability at the crest of an axial wave.

Other examples of atomization will also be briefly presented.

2.2. Generation of droplets through vibrations normal to the liquid layer

In the first example mentioned above, the acceleration field being studied is given and is due to the *forced vibration of a container* (Lang 1962; Parker *et al.* 2008; Boukra *et al.* 2009).

It is proposed that droplets should be generated through ultrasonic vibrations in a container, with the vibrations being normal to the liquid layer of thickness h. The resulting instability is called a parametric instability or *Faraday instability*. For certain frequencies and amplitudes of oscillation, the liquid surface is significantly deformed and reveals nonlinear structures of filaments that are more or less intertwined, giving rise to even more droplets, as shown in Figure 2.2.

The treatment summarized below was used by Benjamin and Ursell (1954), and used again by Marie Lalo (2006). As shown in Figure 2.3, we consider the situation of a liquid layer in a horizontal container, which is subject to a vertical forced oscillation.

Figure 2.2. *The atomization of a solution of distilled water and 80% glycerin, at a frequency of 20 Hz, from a photo taken by Goodridge (Goodridge* et al. *1997). We note that the surface is heavily disturbed with filaments, giving rise to droplets*

Let us write the equations for motion in the reference frame associated with the container:

$$\nabla \cdot \mathbf{V} = 0, \ \mathbf{F} = \rho\left(g - f'\cos\omega t\right)\mathbf{e}_z, \ \frac{\partial \mathbf{V}}{\partial t} + \mathbf{V} \cdot \nabla \otimes \mathbf{V} + \frac{1}{\rho}\nabla p + \frac{1}{\rho}\mathbf{F} = \mathbf{0} \qquad [2.2]$$

It is hypothesized that there is a velocity potential ϕ.

The boundary conditions after linearization are the following, at z=0:

$$\left.\frac{\partial \phi}{\partial t}\right|_{z=0} + \frac{\sigma}{\rho}\left(\frac{\partial^2 \zeta}{\partial x^2} + \frac{\partial^2 \zeta}{\partial y^2}\right) - \left(g - f'\cos \omega t\right) z = 0 \,, \quad \frac{\partial \zeta}{\partial t} = \frac{\partial \phi}{\partial z} \qquad [2.3]$$

Let $k_m^{\ 2}$ be the eigenvalues for the Laplacian of the surface $\nabla_{//}^{\ 2} = \frac{\partial^2}{\partial x^2} + \frac{\partial^2}{\partial y^2}$.

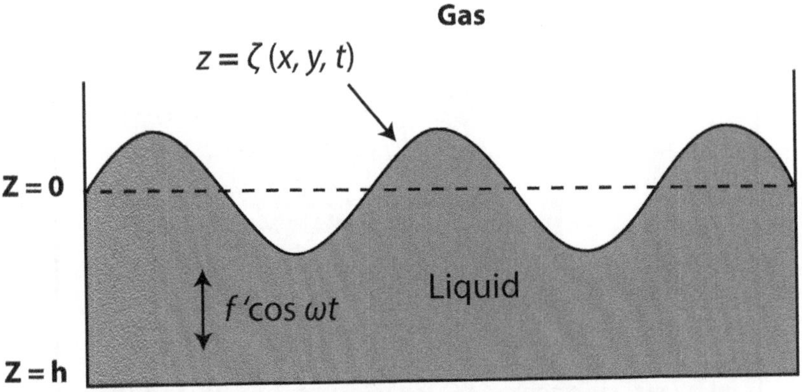

Figure 2.3. *Liquid mass excited by ultrasonic oscillation. The Oz axis is oriented along the descending vertical. The xOy plane is horizontal. For a color version of this figure, see www.iste.co.uk/prudhomme/fluid2.zip*

The independent eigenfunctions $S_m(x, y)$ are such that

$$\left(\frac{\partial^2}{\partial x^2} + \frac{\partial^2}{\partial y^2} + k_m^{\ 2}\right) S_m(x, y) = 0$$

Decomposing on the S_m base gives:

$$\zeta(x, y, t) = \sum_0^{\infty} a_{m(t)} S_m(x, y)$$

The following equation is obtained:

$$\frac{d^2 a_m}{dt^2} + k_m \text{th}(k_m h)\left(\frac{k_m{}^2 \sigma}{\rho} + g - f'\cos\omega t\right)a_m = 0$$

Writing: $T = \omega t/2$ and

$$\begin{cases} p_m = \dfrac{4k_m \tanh(k_m h)}{\omega^2}\left(\dfrac{k_m^2 \sigma}{\rho} + g\right) \approx \text{frequency} \\ q_m = \dfrac{2k_m f' \tanh(k_m h)}{\omega^2} \approx \text{oscillation amplitude} \end{cases}$$

The Mathieu–Hill (Sindayihebura 1995) equation can be derived:

$$\frac{d^2 a_m}{dT^2} + (p_m - 2q_m \cos 2T)a_m = 0$$

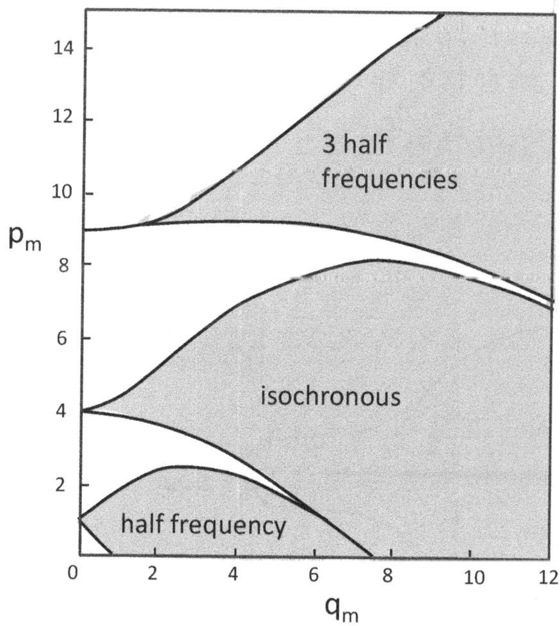

Figure 2.4. *Stability diagram of a liquid layer subject to forced oscillation: separated instability zones can be observed*

Finally, we find that $a_m(T)=e^{i\mu T}\Theta(T)$, $\mu=\mu(p_m,q_m)$, where Θ is a periodic function of *T*.

The stability diagram is presented in Figure 2.4.

The stability of the free surface of a liquid subject to forced oscillation depends on the excitation that is imposed (amplitude of the acceleration *f'* and pulsation ω) as well as on the properties of the liquid (surface tension and density).

2.3. Rayleigh–Taylor instability at the crest of an axial wave

In this second example, the acceleration field under study arises from friction forces produced by the flow of a gaseous jet over a liquid wave, initially created by this same flow on the surface of the liquid, which is planar in the reference frame. This is a *phenomenological study*.

The flow structure studied is presented in Figure 2.5.

The superimposed injection of fluids is such that the *ratio of the momentum* of the gas and the liquid is large.

$$M=\rho_G U_G^2/\rho_L U_L^2 >> 1$$

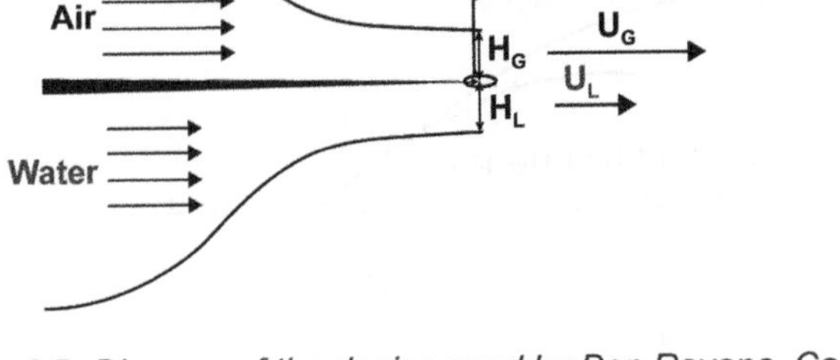

Figure 2.5. *Diagram of the device used by Ben Rayana, Cartellier and Hopfinger (Ben Rayana* et al. *2006). δ_G is the vorticity thickness. For a color version of this figure, see www.iste.co.uk/prudhomme/fluid2.zip*

According to Ben Rayana *et al.*, the events occur as described below.

First, axial waves are formed (waves aligned perpendicular to the direction of flow) following a linear Kelvin–Helmholtz process. Note that the wavelength can be written as:

$$\lambda_{axi} = C_{axi}\sqrt{\rho_L/\rho_G}\,\delta_G$$

These waves are amplified in a nonlinear manner and are themselves subject to an acceleration field at their crest.

This acceleration field is not taken into account in the linearized study of axial waves. To study this new instability, we must start from the crest of the axial wave, subject to acceleration due to the action of the external gas (Figure 2.6) and equal to

$$a=(C_D\,\rho_G/2b\,\rho_L)(U_G-U_C)^2$$

where b is the thickness of the crest (a clearly definite fraction of the wavelength λ_{axi}) and C_D is the drag coefficient.

At this level, a Rayleigh–Taylor type of instability is produced between two superimposed fluids (Varga *et al.* 2003). The acceleration due to gravity g must simply be replaced by the acceleration a.

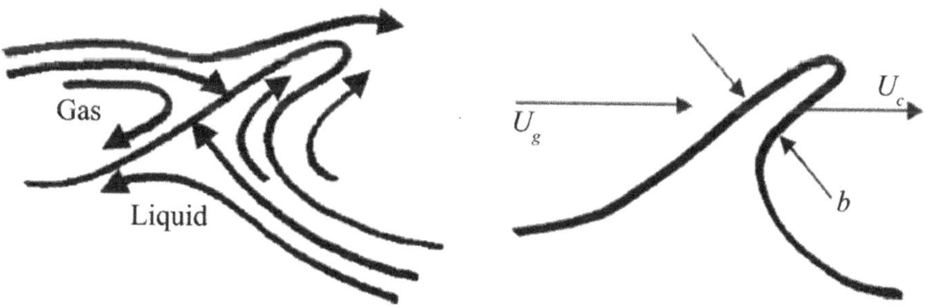

Figure 2.6. *Study of the crest of the axial wave resulting from a convective Kelvin–Helmholtz instability. The convection velocity is (Dimotakis 1986):*
$U_C = U_G\left(1 - U_L/U_G\sqrt{\rho_L/\rho_G}\right)/\left(1+\sqrt{\rho_L/\rho_G}\right)$

It is thus found that:

$$\lambda_{RT} = 2\pi\sqrt{3\sigma/a\rho_L}$$

Finally, from the relations given above:

$$\lambda_{RT} = 2\pi\sqrt{\frac{3\alpha C_{axi}}{C_D}\delta_G}\left(\frac{\rho_L}{\rho_G}\right)^{1/4}\left[\frac{\rho_G\left(U_G - U_C\right)^2\delta_G}{\sigma}\right]^{-1/2}$$

Accepting the hypothesis *M>>1* and writing:

$$Wec_0 = \frac{\rho_G\left(U_G - U_C\right)^2\delta_G}{\sigma}$$

we have:

$$\lambda_{RT} = 2\pi\sqrt{\frac{3\alpha C_{axi}}{C_D}\delta_G}\left(\frac{\rho_L}{\rho_G}\right)^{1/4} Wec_0^{-1/2}$$

Figure 2.7 shows the filaments attached to the crest of the Kelvin–Helmholtz wave, which are caused by the Rayleigh–Taylor instability. It can also be seen that droplets detach from these filaments.

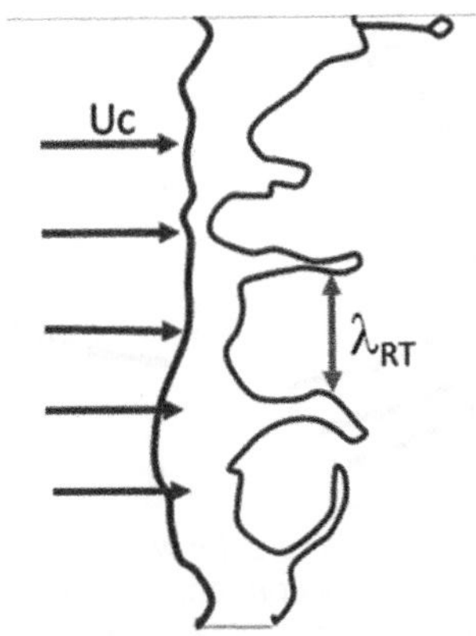

Figure 2.7. *Transverse waves with the wavelength* $\lambda_{TR} \propto U_G^{-1/4}\left(U_G - U_C\right)^{-1}$ *(seen from above, from Ben Rayana et al. 2006). For a color version of this figure, see www.iste.co.uk/prudhomme/fluid2.zip*

2.3.1. *Size distribution of the drops*

It can be experimentally verified that the average size of the drops is proportional to λ_{RT}.

REMINDER (PRUD'HOMME 2014).–

Continuous distribution: the statistical sizes of the drops are defined from the continuous distribution function *f(D)*.

$$D_{pq} = \left(\int_0^\infty D^p f(D)\, dD \Big/ \int_0^\infty D^q f(D)\, dD \right)^{1/p-q}$$

Discrete distribution (finite number of drops):

$$D_{pq} = \left(\sum_1^N d_i^p \Big/ \sum_1^N d_i^q \right)^{1/p-q}$$

We generally consider the mean diameter D_{10} and the Sauter mean diameter D_{32} which is equal to the sum of the volumes divided by the sum of the areas.

NOTE.–

It must be noted that Lasheras and Hopfinger (2000) presented the topology of mechanisms for the atomization of liquid jets. The typical regime of the LPRE (Liquid-Propellant Rocket Engine) coaxial injector is a filament regime (fiber-type atomization), which has been presented here. This regime is seen when the Weber number for the liquid drops, We(D_L), is greater than a threshold that lies between 100 and 1,000 (inclusive) dependent on the liquid Re (see Figure 2.8). This occupies a small domain in the stability diagram, where:

$$We = \rho_G U_G D_L / \sigma, \ \mathrm{Re}_L = U_L D_L / \nu_L, \ M = \rho_G U_G^2 / \rho_L U_L^2$$

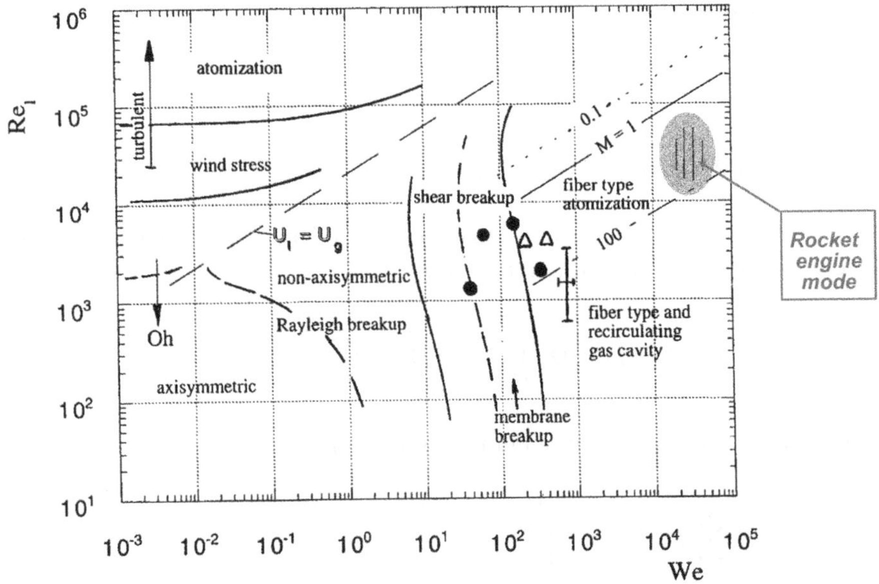

Figure 2.8. *Rupture regimes in space for the parameters Re_L/We. The lines with constant M are calculated for water–air and D_L=7 mm. For a color version of this figure, see www.iste.co.uk/prudhomme/fluid2.zip*

2.4. Recent work

Stochastic models were proposed by Gorokhovski *et al.* (2009).

Research is ongoing at ONERA using an installation and specific diagnostics in order to determine the appropriate laws for the mean diameters and the size distribution of droplets. It is especially crucial to have correct formulae for the role of the ratio of densities when these densities are very different (Porcheron *et al.* 2002). Research has been published in 2016 (Fdida *et al.* 2016), 2017 (Fdida *et al.* 2017) and 2019 (Fdida *et al.* 2019).

2.5. Conclusion

Fluids in motion present interfacial instabilities that cannot always be shown by studying small, linearized disturbances. Thus, between two parallel jets with very different velocities, if the linear study reveals a

Kelvin–Helmholtz instability, the later development of this primary instability must be taken into account to explain the appearance of another instability. By separately considering the acceleration field created at the crest of the primary wave, we can explain this new instability that is similar to a Rayleigh–Taylor instability.

It is probable that a nonlinearized 3D calculation would show these two instabilities from the beginning; however, to the best of our knowledge, no such calculation has been published so far (Fuster *et al.* 2013).

The numerical solution was carried out in 2D by Fuster *et al.*, who compared it with the analytical solution and experimental results. The DNS codes that currently exist should allow a 3D solution.

This is a significant challenge as the liquid structures that appear are a stage of atomization into droplets (Villermaux *et al.* 2004). Furthermore, *the size of the droplets* plays an important role in designing engines with efficient combustion. New experiments are being developed in ONERA to study this phenomenon (Mauriot *et al.* 2015). It is essential to experimentally verify the formulae in the literature for the average size and the size distribution of droplets. Indeed, significant variations can be seen between many empirical and semi-empirical theoretical formulae.

2.6. References

Antkowiak, A., Bremond, N., Le Dizès, S., Villermaux, E. (2007). Short-term dynamics of a density interface following an impact. *Journal of Fluid Mechanics*, 577, 241–250.

Azzopardi, B.J. (1988). The role of drops in annular gas-liquid flow: Drop sizes and velocities. *Japanese Journal of Multiphase Flow*, 2(1), 15–35.

Ben Rayana, F., Cartellier, A., Hopfinger, E.J. (2006). Assisted atomization of a liquid layer: Investigation of the parameters affecting the mean drop size prediction. *ICLASS06-190*, 27 August–1 September, Kyoto.

Benjamin, T.B. and Ursell, F. (1954). The stability of the plane free surface of a liquid in vertical periodic motion. *Proceedings of the Royal Society of London*, 225(1163), 505–515.

Boukra, M., Cartellier, A., Ducasse, E., Gajan, P., Lalo, M., Noel, T., Strzelecki, A. (2009). Use of Faraday instabilities to enhance fuel pulverisation in air-blast atomisers. Application de méthodes actives pour l'amélioration de la pulvérisation dans un injecteur aérodynamique. *Comptes rendus mécanique*, 337(6–7), 492–503.

Chandrasekhar, S. (1961). *Hydrodynamic and Hydromagnetic Stability*. Clarendon Press, Oxford University Press, Oxford.

Dimotakis, P.E. (1986). Two-dimensional shear-layer entrainment. *AIAA Journal,* 24(11), 1791–1796.

Fdida, N., Mauriot, Y., Vingert, L., Ristori, A., Godel, S., Théron, M. (2016). Caractérisation de l'atomisation d'un jet d'oxygène liquide assisté par azote gazeux au moyen de diagnostics optique. *Congrès francophone des techniques laser (CFTL)*, 13–16 September, Toulouse, France.

Fdida, N., Mauriot, Y., Vingert, L., Nugue, M. (2017). Granulométrie par imagerie d'un écoulement diphasique cryogénique. *Instrumentation, mesure, métrologie*, Lavoisier, 16(16), 229–253.

Fdida, N., Mauriot, Y., Vingert, L., Ristori, A., Théron, M. (2019). Characterizing primary atomization of cryogenic LOX/Nitrogen and LOX/Helium sprays by visualizations coupled to Phase Doppler Interferometry, *Acta Astronautica*, 164, 458–465.

Fuster, D., Matas, J.P., Marty, S., Popinet, S., Hoepffner, J., Cartellier, A., Zaleski, S. (2013). Instability regimes in the primary breakup region of planar coflowing sheets. *Journal of Fluid Mechanics*, 736, 150–176.

Goodridge, C.L., Tao Shi, W., Hentschel, H.E., Lathrop, D.P. (1997). Viscous effects in droplet-ejecting capillary waves. *Physical Review E*, 56, 472–475.

Gorokhovski, M., Jouanguy, J., Chtab-Desportes, A. (2009). Stochastic model of the near-to-injector spray formation assisted by a high-speed coaxial gas jet. *Fluid Dynamics Research*, 41(3).

Inoue, C., Izato, Y., Miyake, A., Villermaux, E. (2017). Direct self-sustained fragmentation cascade of reactive droplets. *Physical Review Letters*, 118(7), 074502.

Lalo M. (2006). Atomisation d'un film mince liquide par action combinée des instabilités de Kelvin–Helmholtz et de Faraday. Application aux injecteurs aérodynamiques des turbomachines aéronautiques. ENSAE-ONERA thesis, Toulouse, 21/12/2006.

Lamb, H. (1945). *Hydrodynamics*. Cambridge University Press, Cambridge.

Lang, R.J. (1962). Ultrasonic atomization of liquids. *The Journal of the Acoustical Society of America*, 34(1), 6–8.

Lasheras, J.C. and Hopfinger, E.J. (2000). Liquid jet instability and atomization in a coaxial gas stream. *Annual Review of Fluid Mechanics*, 32, 275–308.

Marmottant, P. and Villermaux, E. (2004). Fragmentation of stretched liquid ligaments. *Physics of Fluids*, 16(8), 2732–2741.

Mauriot, Y., Vingert, L., Dorey, L.-H. (2015). Characterizing coaxial primary atomization without combustion on the Mascotte test bench. REST Workshop – Atomisation ONERA, 9 June.

Nicoli, C., Haldenwang, P. and Denet, B. (2003), Combustion of gaseous coflow jets. *Combustion Science and Technology*, 175, 1143–1163.

Niemeyer, J. (2015). Rayleigh–Taylor instability [Online]. Available at: https://youtu.be/yabqo7VFTYs.

Parker, S.H., Suh, H.K., Lee, C.S. (2008). Effect of cavitating flow on the flow and fuel atomization characteristics of biodiesel and diesel fuels. *Energy and Fuels*, 22, 605–613.

Porcheron, E., Carreau, J.L., Prevost, L., Le Visage, D., Roger, F. (2002). Effect of injection gas density on coaxial liquid jet atomization. *Atomization and Spray*, 12, 209–227.

Prud'homme, R. (2014). *Flows and Chemical Reactions in Heterogeneous Mixtures*. ISTE Ltd, London, and John Wiley & Sons, New York.

Sindayihebura, D. (1995). Pulvérisation ultrasonique des films liquides de faible épaisseur. PhD thesis, UCL, Louvain.

Varga, C.M., Lasheras, J.C., Hopfinger, E.J. (2003). Initial breakup of a small-diameter liquid jet by a high-speed gas stream. *Journal of Fluid Mechanics*, 497, 405–434.

Villermaux, E., Marmottant, P., Duplat, J. (2004). Ligament-mediated spray formation. *Physical Review Letters*, 92(7), 074501.1–074501.4.

3

Numerical Simulation of Pipes with an Abrupt Contraction Using OpenFOAM

The simulation engine system is an important challenge in the automotive industry as it makes it possible to speed up the design process and limit testing on physical prototypes. The AMESim platform enables access to libraries of models for simulating an engine or a vehicle. Modeling approaches (0D/1D) do not however equal the precision of the 3D codes. Hence, this study is on OpenFOAM (simulation software CFD 3D) in order to reduce the problem to 1D by bringing in new perspectives of computing time and prediction quality.

3.1. Introduction

The development of computational fluid dynamics (CFD) of methods for industrial use was mainly guided by the need to carry out simulations on a 3D geometric model with complex physical behavior. The finite volume method is usually the preferred one for industrial CFD as it is relatively cheaper and lends itself well to solving equations associated with complex flows. Variables such as velocity and pressure are usually colocated, i.e. they are defined over the same set of discrete meshes (the center of the cells).

The presence of discontinuities in the flow (such as shocks and contact surfaces) at high, compressible velocities requires numerical schemes that can capture these characteristics while avoiding false oscillations. Notably, efficient numerical methods that generate exact, non-oscillatory solutions. Methods that generally involve a Riemann solver or a Jacobian evaluation

Chapter written by Tarik CHAKKOUR.

make these difficult to implement in a polyhedric domain, for example. We propose the numerical schemes developed by Nessyahu and Tadmor (1990) as a generalization of the second order of the Lax–Friedrichs scheme. The method was developed for multidimensional systems and made up of non-structured meshes, mainly from triangular, tetrahedral, quadrilateral and hexahedral meshes.

The aim of this chapter is to develop and study abrupt contractions in pipes using OpenFOAM (2016), an open-source CFD-3D simulation software. Since modeling approaches using the AMESim 1D CFD do not match the precision of 3D codes, we will carry out a complete study of the orifices in 3D. These orifices are tubes that make up evacuation pipes in an engine. Although the AMESim library makes it possible to observe results that are quite reliable in terms of the resolution of waves and shocks, it does not allow the simulation of 3D gas flows in more complicated systems like the evacuation collectors. Further, engine designers must take into account the 3D flow of gas within the collectors to predict a better performance. Hence, constructing models using a CFD-3D is needed to re-examine the 1D problem.

There are several studies (Guo *et al.* 2001; Satish *et al.* 2013) of sudden contraction models. The paper by Satish *et al.* (2013) describes the 2D behavior of fluid flow in a pipe with a sudden contraction in the Fluent CFD software. The numerical results generated were validated through a comparison with experimental data. The article by Guo *et al.* (2001) studies a turbulent flow in a pipe followed by a sudden contraction. The transient RANS approach and the k-ϵ model are implemented to demonstrate the instability of the flow over certain zones.

3.2. Modeling an abrupt contraction in a pipe

3.2.1. *Euler equations*

In this section, we will determine the equations that govern the flow dynamics of a non-viscous, compressible fluid in a 3D planar channel. The aim here is to validate our framework using the OpenFOAM numerical simulation. Our description is inspired by the Euler equations and based on a numerical scheme using Kurganov's and Tadmor's finite volumes (Kurganov and Petrova 2000; Kurganov and Tadmor 2000).

The Euler equations for gas dynamics make it possible to describe the conservation of mass, impulsion and energy in a local form as a function of density ρ, velocity $\mathbf{u}$, pressure p and internal energy e. The total energy E is the sum of the internal and kinetic energies. In the case of a perfect polytropic gas, the energy E, the pressure p and the internal energy e are given by the following relations:

$$\begin{cases} E = \frac{1}{2}\|\mathbf{u}\|^2 + e, \\ p = \rho RT, \\ e = c_v \rho T. \end{cases} \qquad [3.1]$$

The conservation of mass is formulated by the following equation:

$$\frac{\partial \rho}{\partial t} + \nabla.[\mathbf{u}\rho] = 0. \qquad [3.2]$$

The conservation of momentum is formulated by the following equation:

$$\frac{\partial \rho \mathbf{u}}{\partial t} + \nabla.[\mathbf{u}(\rho\mathbf{u})] + \nabla p = 0. \qquad [3.3]$$

The conservation of energy is defined by the following equation:

$$\frac{\partial \rho E}{\partial t} + \nabla.[\mathbf{u}(\rho E)] + \nabla.[\mathbf{u}p] = 0. \qquad [3.4]$$

The above three equations are reduced to only a single hyperbolic equation that makes it possible to conserve physical quantities, called the Euler equation:

$$\frac{\partial w}{\partial t} + \nabla.[F(w)] = 0, \qquad [3.5]$$

where w is the conservative vector and F is the numerical flow, which are defined, respectively, by:

$$w = \begin{pmatrix} \rho \\ \rho\mathbf{u} \\ \rho E \end{pmatrix}, \quad F(w) = \begin{pmatrix} \mathbf{u}\rho \\ \mathbf{u}(\rho\mathbf{u}) + pI \\ \mathbf{u}(\rho E + p) \end{pmatrix}. \qquad [3.6]$$

The standard integral form of the Euler equation over a control volume V_P, which is given by the center P, is described in the following form:

$$\int_t^{t+\Delta t} \left[\frac{d}{dt} \int_{V_P} w \, \mathrm{d}V + \int_{V_P} \nabla.[F(w)] \, \mathrm{d}V \right] \mathrm{d}t = 0. \qquad [3.7]$$

3.2.2. *Stability of the solver*

The OpenFOAM software distribution offers several solvers related to compressible flows, with each one being adapted to a specific type of compressible flow. The **rhopSonicFoam** solver is used as the base for a compressible microflow solver, as it was partially redeveloped for the analysis of flows in applications involving hypersonic aerodynamics. The **rhopSonicFoam** solver is an application for pressure, based on the density of compressible laminar flows. It solves equations that govern continuity, momentum (written in terms of flow mass) and total energy. These equations are solved in a distinct way, followed by a pressure correction loop (the **PISO** algorithm described in Dainese (1994)).

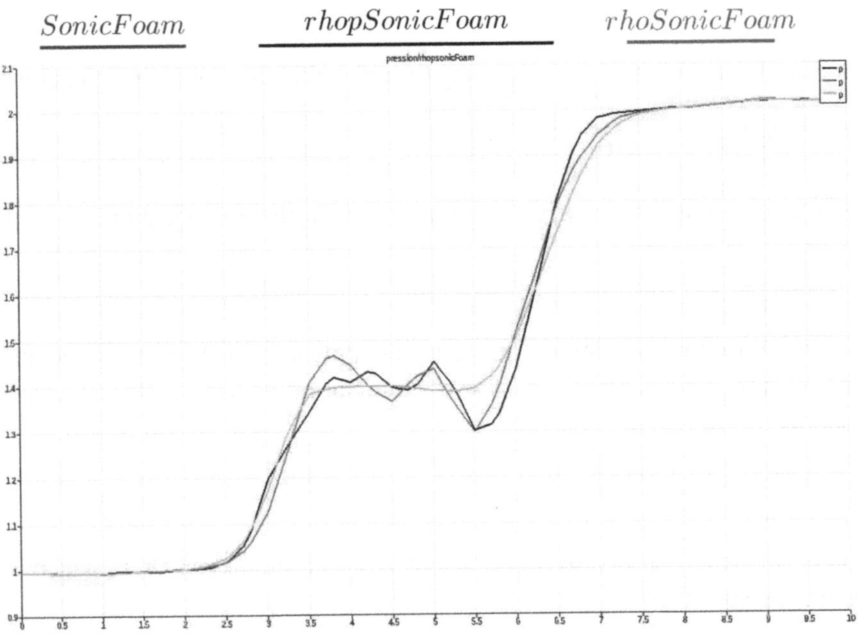

Figure 3.1. *Comparison of the three solvers. For a color version of this figure, see www.iste.co.uk/prudhomme/fluid2.zip*

Both the **rhoSonicFoam** and **rhopSonicFoam** solvers are based on the pressure equation for transsonic problems, including for regions with a low Mach number. On the other hand, **SonicFoam** is a solver that solves Navier–Stokes equations using the **PISO** algorithm. If viscosity is zero, then

it solves Euler equations for a compressible, non-viscous fluid. The **SonicFoam** solver simply calculates internal energy e (thermal energy). Kinetic energy is neglected and not calculated.

Now, we wish to identify the most stable solver in order to model a compressible, non-viscous flow with Euler equations. In order to do this, we configure the problem of a tube experiencing a shock using the limiting conditions for a blocked pipe. Figure 3.1 presents the evolution of a shockwave with variable pressure for the three solvers: **SonicFoam**, **rhoSonicFoam** and **rhopSonicFoam**. This simulation is carried out for the same time and space step. As the pressure curve for the **SonicFoam** scheme is given in blue, we conclude that it is the most stable solver.

3.2.3. *Introducing the model*

Let S be a given section. We define a contraction model when the variation of the section is negative, $dS < 0$, and we define an enlargement model when the variation of the section is positive $dS > 0$. Our physical domain is a horizontal pipe, $0.2m$ long, subject to a contraction at $x = 0.1m$, for which the left half has a diameter of $0.04m$ and the right half has a diameter of $0.03m$. It is specified that the quotient of the square of the right half, and the left one, must be greater than $0.5m$.

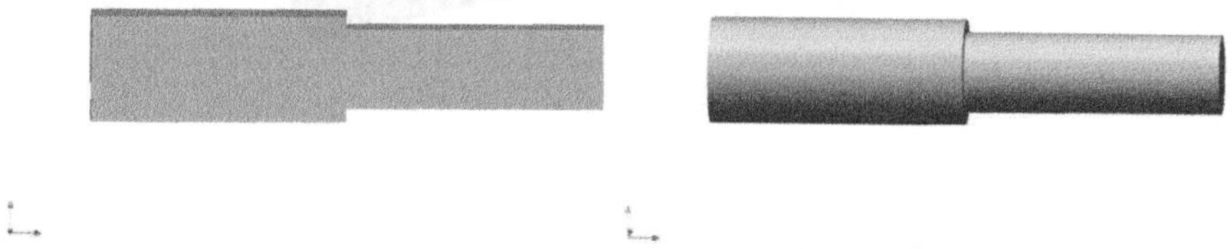

(a) (b)

Figure 3.2. *(a) Cartesian domain; (b) cylindrical domain*

Two computational domains are considered: Cartesian and cylindrical domains. This is done to understand whether or not the geometry of the domain influences the nature of the mean flow. Two different types of meshes are then investigated (relaxed and refined) over these two domains to find out whether or not our meshing is convergent with respect to our numerical simulation. This is a uniform mesh across the three directions in space: $\Delta x = \Delta y = \Delta z$ with varied lengths over this domain.

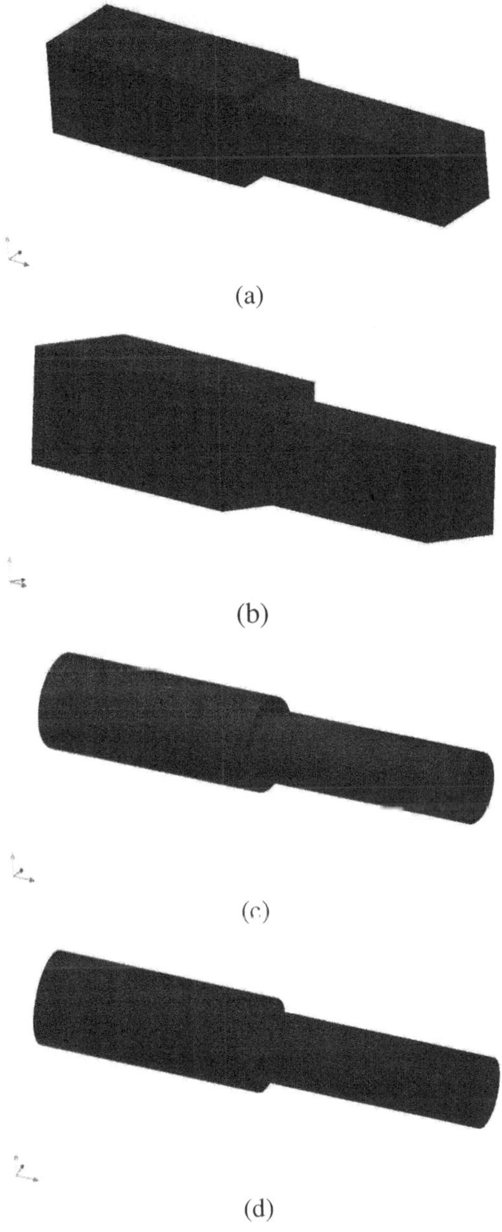

Figure 3.3. *(a) Relaxed Cartesian mesh; (b) refined Cartesian mesh; (c) relaxed cylindrical mesh; (d) refined cylindrical mesh. For a color version of this figure, see www.iste.co.uk/prudhomme/fluid2.zip*

The computational domain is subdivided into several control volumes of any shape and size. These cells must cover entirely the domain in order to provide a good approximation in terms of the geometry of the flow. For a relaxed mesh, the small section is discretized with a number of $n_x \times n_y \times n_z$ meshes and the large section is discretized with a number of $m_x \times m_y \times m_z$ meshes at the location $(\vec{e_x}, \vec{e_y}, \vec{e_z})$. We denote the time step by Δt. More formally:

$$\begin{cases} Relaxed\ mesh, \\ \Delta x = \Delta y = \Delta z = 0.00125m, \\ n_x = 12, n_y = 12, n_z = 80, \\ m_x = 16, m_y = 16, m_z = 80, \\ \Delta t = 2.68 \times 10^{-6}. \end{cases}$$

For a refined mesh, the small section is discretized with a number of $n_x \times n_y \times n_z$ meshes and the large section is discretized with a number of $m_x \times m_y \times m_z$ meshes at the location $(\vec{e_x}, \vec{e_y}, \vec{e_z})$:

$$\begin{cases} Refined\ mesh, \\ \Delta x = \Delta y = \Delta z = 0.000625m, \\ n_x = 24, n_y = 24, n_z = 160, \\ m_x = 32, m_y = 32, m_z = 160, \\ \Delta t = 1.34 \times 10^{-6}. \end{cases}$$

3.2.4. *Boundary and initial conditions*

The fluid flow must be stationary, circulating from left to right, initially with null velocity. In order for this to happen, the **symmetryPlane** boundary conditions are imposed on the lateral surface such that the flow is unidimensional. Let us differentiate between two types of boundary conditions and initial conditions:

– We enforce a temperature difference ΔT at $57K$ with a null pressure gradient vector and with input velocity fixed at $50m/s$, and a null input pressure gradient vector. At the outlet, we impose a velocity gradient vector and a null temperature gradient vector. These boundary conditions and initial conditions are presented in Figure 3.4.

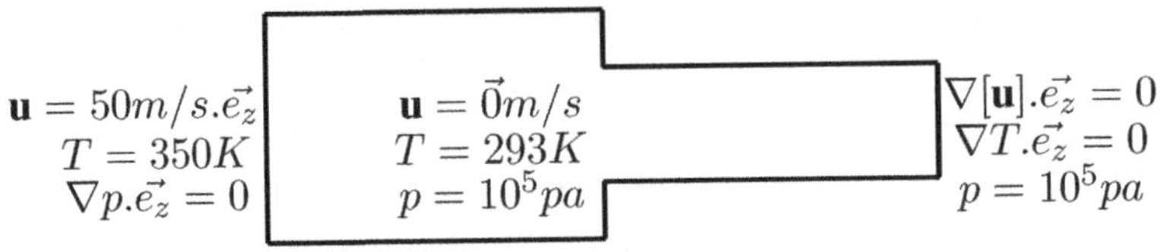

Figure 3.4. *Boundary and initial conditions I*

– In this case, we define stationary initial conditions. In order to do this, we impose a temperature difference ΔT of $300K$ with a constant output pressure fixed at $p = 10^5 pa$ and a relatively high input velocity. These boundary and initial conditions are presented in Figure 3.5.

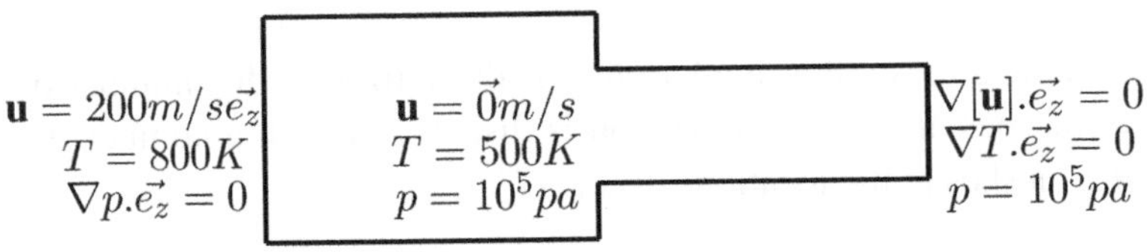

Figure 3.5. *Boundary conditions and initial conditions II*

Let S_i be the area of each cell C with a section that is perpendicular to the direction of flow, $\vec{e_z}$, in our pipe. If v is a variable presenting a physical quantity, the mean variable $\bar{v}$ is defined by the mean v per unit area over each cell C as given below:

$$\bar{v} = \frac{\sum_{i \in C} v_{z,i} S_i}{\sum_{i \in C} S_i}. \qquad [3.8]$$

Relation [3.8] is used to define the mean velocity $\bar{\mathbf{u}}_z$, the average density $\bar{\rho}_z$ and the average pressure $\bar{p}_z$, respectively, from the velocity $\mathbf{u}_z$, the density ρ_z and the pressure p_z. Let us recall that the flow Q is defined by the density integral multiplied by the velocity in the direction of flow $\vec{e_z}$ over the surface element dS

$$Q = \int_S \rho \mathbf{u}_z. \qquad [3.9]$$

From definition [3.9] for the mean flow, we can differentiate between two expressions, D and d. The first flow rate, D, is taken by considering fluctuations and is defined from the mean velocity multiplied by the density $\overline{\rho_z \mathbf{u}_z}$ using:

$$D = \overline{\rho_z \mathbf{u}_z} \times S. \qquad [3.10]$$

We will now validate the expression for the average flow rate D defined in relation [3.10]. The mean flow rate may be defined as:

$$\bar{Q} = \int_S \overline{\rho \mathbf{u}_z}. \qquad [3.11]$$

Integrating over a constant section S gives expression [3.10] for the mean flow rate D. We will now validate the expression for the second flow rate d, which is defined as the product of the mean density $\bar{\rho}_z$, the mean velocity $\bar{\mathbf{u}}_z$ and the surface S:

$$d = \bar{\rho}_z \bar{\mathbf{u}}_z \times S. \qquad [3.12]$$

It must be remembered that due to the conservation of mass, the flow rate of a fluid in a pipe is constant. If we decompose the instantaneous velocity $\mathbf{u}_z$ in the form of a sum of the mean velocity $\bar{\mathbf{u}}_z$, and the fluctuating velocity $\underline{\mathbf{u}_z}$ such that the mean of the fluctuation velocity is null, i.e. $\mathbf{u}_z = \bar{\mathbf{u}}_z + \underline{\mathbf{u}_z}$, we obtain:

$$\bar{Q} = \int_S (\bar{\rho} + \underline{\rho})(\bar{\mathbf{u}}_z + \underline{\mathbf{u}_z}). \qquad [3.13]$$

We then simplify the expression for the mean flow rate, $\bar{Q}$, defined in relation [3.13] to obtain the following relation:

$$\bar{Q} = \int_S \bar{\rho}\bar{\mathbf{u}}_z + \int_S (\underline{\rho}.\bar{\mathbf{u}}_z + \underline{\rho}.\underline{\mathbf{u}_z} + \bar{\rho}.\underline{\mathbf{u}_z}). \qquad [3.14]$$

The first integral is simplified as the mean values are constant over the section S. The second part of the integral is made null by shifting to the mean of a fluctuating quantity. We then have:

$$\int_S \overline{(\underline{\rho}.\bar{\mathbf{u}}_z + \underline{\rho}.\underline{\mathbf{u}_z} + \bar{\rho}.\underline{\mathbf{u}_z})} = 0, \qquad [3.15]$$

we obtain expression [3.12] for the mean flow rate d over a constant section. It is possible to similarly differentiate between two expressions for mean enthalpy, H and h. The first enthalpy H is considered in fluctuations and is defined from the average of the quantity $E + \frac{p}{\rho}$:

$$H = \overline{E + \frac{p}{\rho}}. \qquad [3.16]$$

This expression for the mean enthalpy H may be simplified in the following form:

$$H = \overline{E} + \frac{\overline{p}}{\rho}. \qquad [3.17]$$

The second enthalpy h is defined by the sum of the total mean energy $\bar{E}$ and the quotient of the mean pressure $\bar{p}$ and the mean density $\bar{\rho}$

$$h = \bar{E} + \frac{\bar{p}}{\bar{\rho}}. \qquad [3.18]$$

3.3. Numerical results

In this section, we wish to evaluate the enthalpy and mean flow rate over time and space through different positions that are close to the change in the section. The conservation of enthalpy h (or H) is equivalent to the conservation of momentum. Thus, the conservation of total energy is equivalent to the conservation of entropy s (or S).

In Figure 3.6, consider the upstream and downstream flow positions numbered 1, 2 and 3. The downstream position represents the position where the fluid flows in the direction of flow. On the other hand, the upstream position is defined by the position against the direction of flow. The upstream positions 1, 2 and 3 are increasingly farther away from the change in the section, in the ratios $3\Delta z$, $2\Delta z$ and Δz, respectively. We will use the same definition for the downstream positions 1, 2 and 3.

Figure 3.7 presents the section of each upstream and downstream position in order to calculate the mean of each of the variables ρ, $\mathbf{u}_z$ and $\rho\mathbf{u}_z$. This makes it possible to provide the mean flow rate D calculated based on the mean of $\overline{\rho\mathbf{u}_z}$. Also, the mean flow rate d is calculated differently by multiplying the mean density over the section $\bar{\rho}$ and the mean velocity $\bar{\mathbf{u}}_z$ and the area of the section.

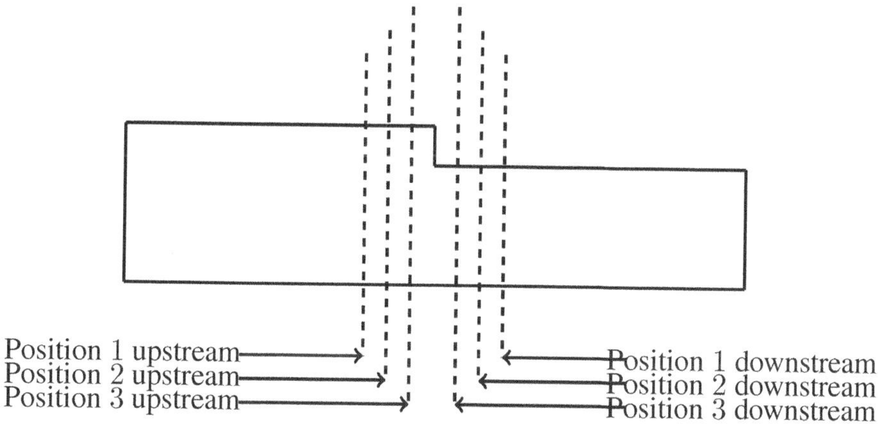

Figure 3.6. *Upstream and downstream positions*

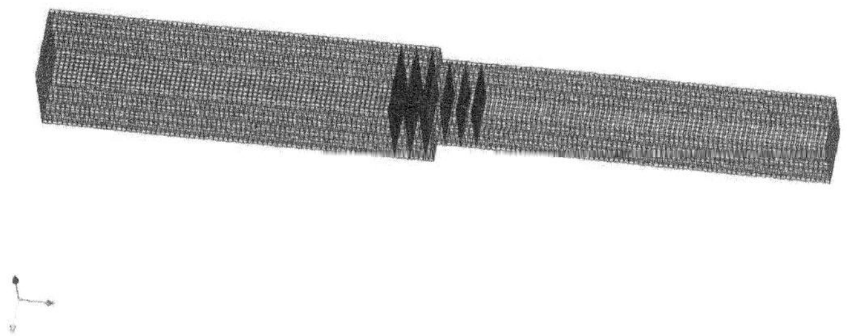

Figure 3.7. *Different sections showing the upstream and downstream positions* 1, 2 *and* 3. *For a color version of this figure, see www.iste.co.uk/prudhomme/fluid2.zip*

3.3.1. *Results with the boundary and initial conditions I*

3.3.1.1. *Variation in mean flow rate and mean enthalpy*

This section will demonstrate how the mean flow rate and mean enthalpy vary over the space defined between the upstream position 1 and the downstream position 1. We will study the variation of this mean flow rate over different temporal iterations. Indeed, we wish to show that the mean flow rate and the mean enthalpy are conserved. In order to do this, we trace the mean flow rate in Figures 3.8a and 3.8b and the mean enthalpy in Figures 3.8c and 3.8d as a function of this defined space. It can be noted that the mass flow rate is only conserved between two sections of the pipe. This law of conservation is defined by the continuity equation. Similarly, the conservation of enthalpy is equivalent to the conservation of momentum. The results of the simulation confirm that there are instants where the mean flow rate (or, respectively, the mean enthalpy) is not at all conserved (or, respectively, is conserved). These instants coincide with the crossing of the wave at the moment when the section changes.

The numerical simulations defined in Figures 3.9a, 3.9b, 3.9c, 3.9d, 3.10a, 3.10b, 3.10c, 3.10d and 3.10e are carried out with respect to a first-order Euler temporal schema. We choose the time step for a relaxed mesh that is double that of a refined mesh in order to carry out an IsoCFL simulation. These numerical results for the mean flow rate over each $4 \times \Delta t$ time step are presented. It is specified that the return wave takes more time than the outward wave. Both the mean flow rates, D and d, defined respectively in relations [3.10] and [3.12], are given in these figures. A comparison of the mean flow rate between Figures 3.10a and 3.10b for position 1 and between Figures 3.10c and 3.10d for position 2 shows that the geometry of the calculation domain does not influence the nature of the mean flow.

For a relaxed Cartesian mesh, we simulate the variation in the mean flow rate over time. It can be seen for the different positions 1, 2 and 3 that the mean flow rate is less two phase as we get closer to the change in section and that it becomes more dephasing as we move further away from the change in section during the stationary state. When the wave moves outward, there are mainly two physical regimes that are established. A stationary regime whose flow rate is null, followed by a transient regime that creates two phases in the upstream and downstream flow rates. When the wave returns, the regime becomes stationary. For the time being, we will study the transient regime.

We choose a simulation interval with the time step Δt in order to estimate the dephasage between the upstream flow and the downstream flow. By default, a relaxed mesh is chosen. For a refined Cartesian mesh, the same behavior is observed for the mean flow rate as for a relaxed mesh. There is simply a difference of 110 time steps. The wave has not yet reached the end of its outward path. At the level of the meshes, however, the schema remains convergent with respect to our numerical simulation. As we move closer to the change in section, no fluctuation in mean flow rate is observed. It can be concluded that this approach is relatively convergent with respect to our mesh.

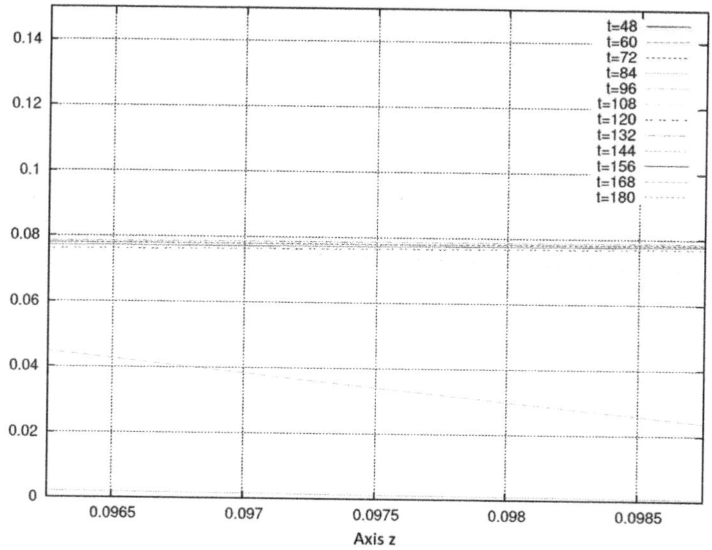
t=48
t=60
t=72
t=84
t=96
t=108
t=120
t=132
t=144
t=156
t=168
t=180
0.14
0.12
0.1
0.08
0.06
0.04
0.02
0
0.0965
0.097
0.0975
0.098
0.0985
Axis z

(a)

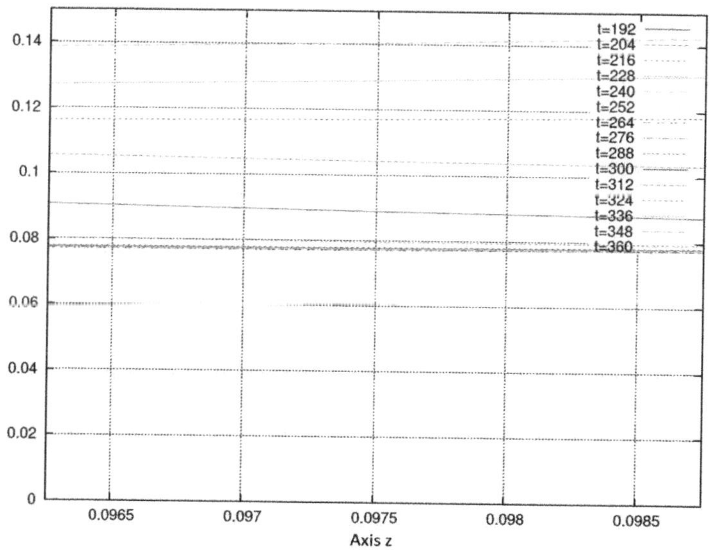
t=192
t=204
t=216
t=228
t=240
t=252
t=264
t=276
t=288
t=300
t=312
t=324
t=336
t=348
t=360
0.14
0.12
0.1
0.08
0.06
0.04
0.02
0
0.0965
0.097
0.0975
0.098
0.0985
Axis z

(b)

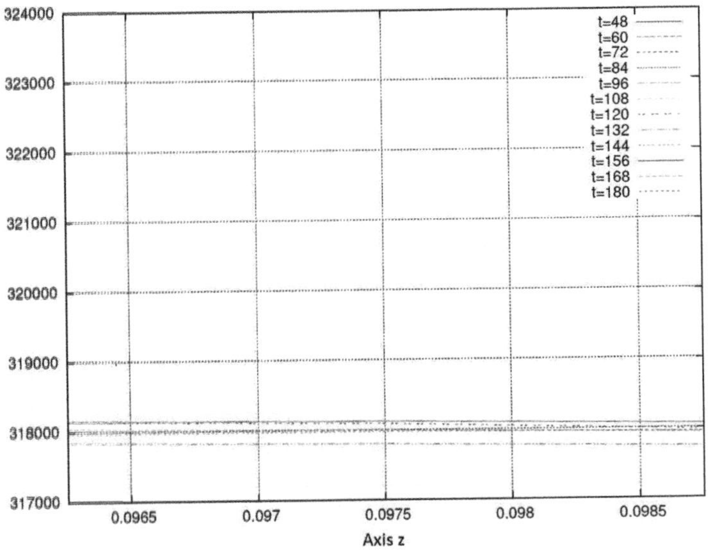

(c)

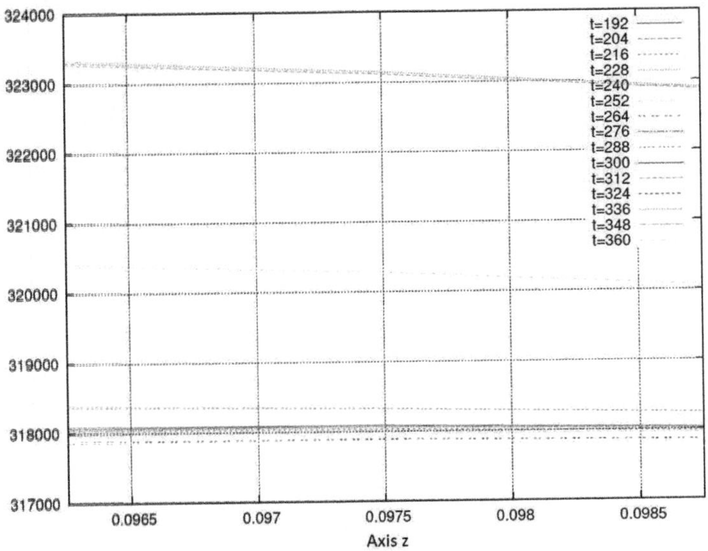

(d)

Figure 3.8. *(a) Mean flow rate over space during the outward movement; (b) mean flow rate over space during the return; (c) mean enthalpy over space during the outward movement; (d) mean enthalpy over space during the return. For a color version of this figure, see www.iste.co.uk/prudhomme/fluid2.zip*

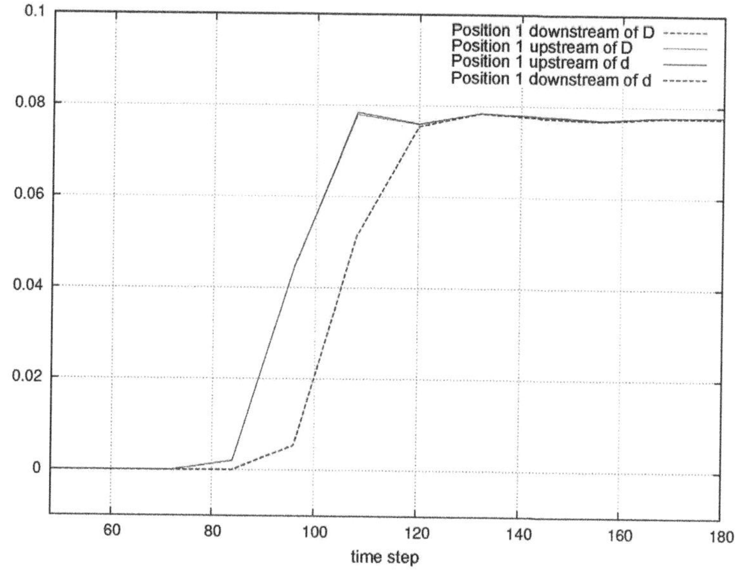
Position 1 downstream of D
Position 1 upstream of D
Position 1 upstream of d
Position 1 downstream of d
0.1
0.08
0.06
0.04
0.02
0
60
80
100
120
140
160
180
time step

(a)

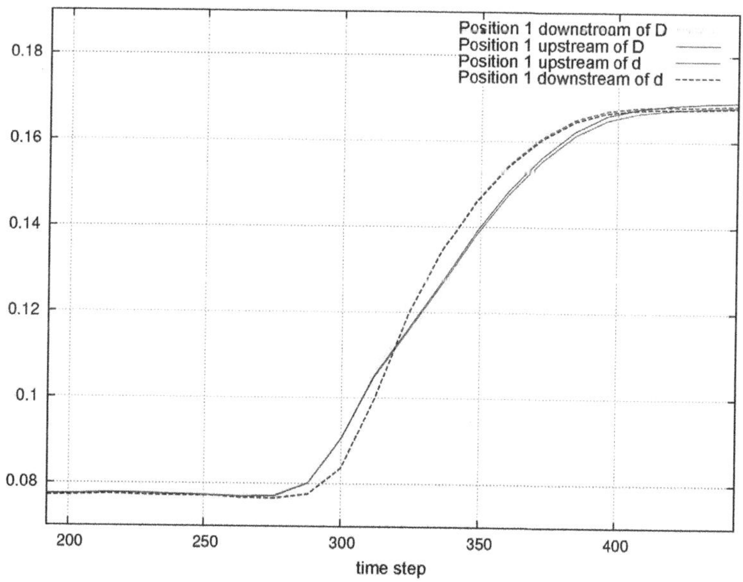
Position 1 downstream of D
Position 1 upstream of D
Position 1 upstream of d
Position 1 downstream of d
0.18
0.16
0.14
0.12
0.1
0.08
200
250
300
350
400
time step

(b)

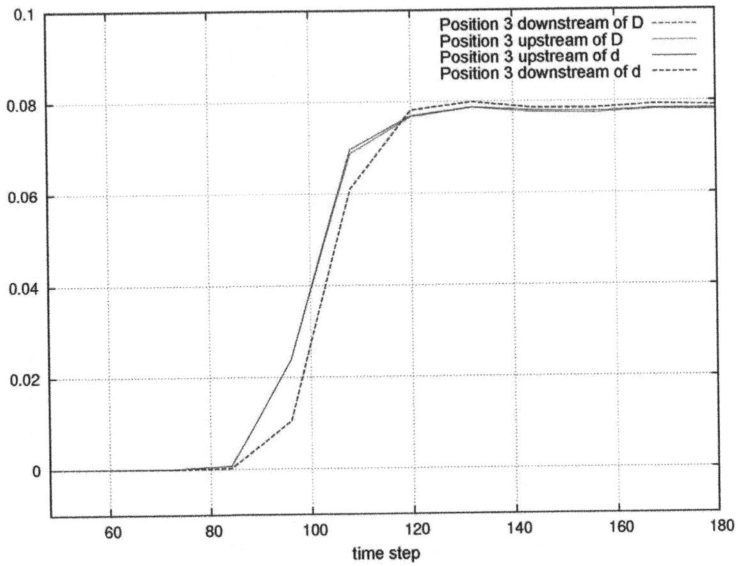

(c)

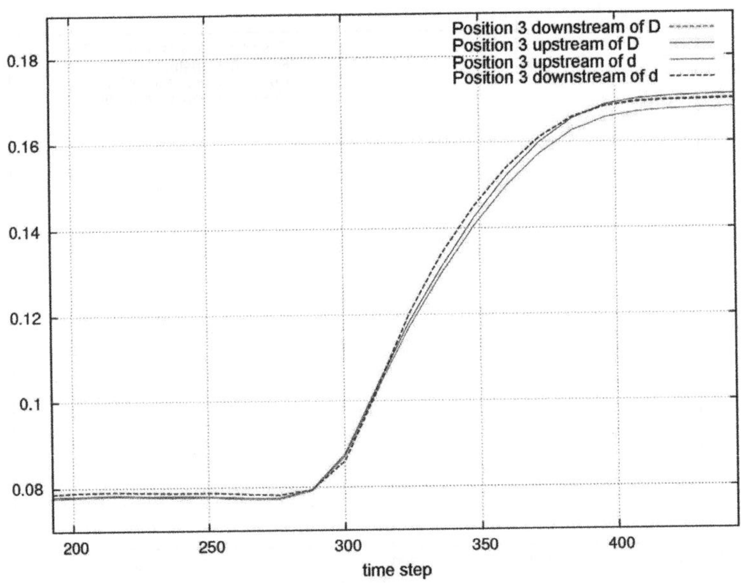

(d)

Figure 3.9. *(a) Mean flow rate for position 1 during the outward movement; (b) mean flow rate for position 1 during the return; (c) mean flow rate for position 3 during the outward movement; (d) mean flow rate for position 3 during the return. For a color version of this figure, see www.iste.co.uk/prudhomme/fluid2.zip*

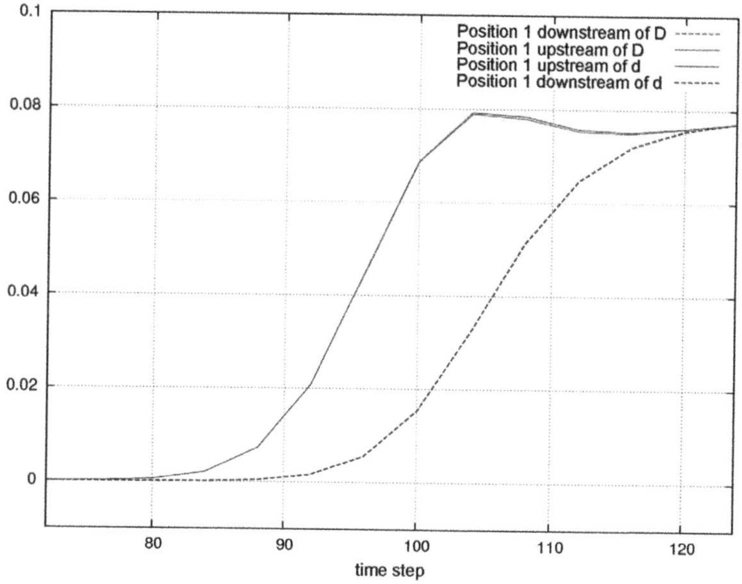
Position 1 downstream of D
Position 1 upstream of D
Position 1 upstream of d
Position 1 downstream of d
0.1
0.08
0.06
0.04
0.02
0
80
90
100
110
120
time step

(a)

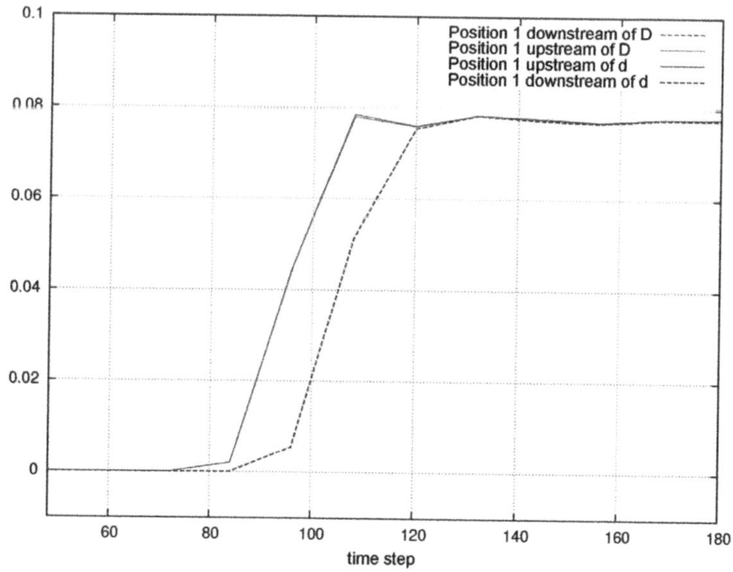
Position 1 downstream of D
Position 1 upstream of D
Position 1 upstream of d
Position 1 downstream of d
0.1
0.08
0.06
0.04
0.02
0
60
80
100
120
140
160
180
time step

(b)

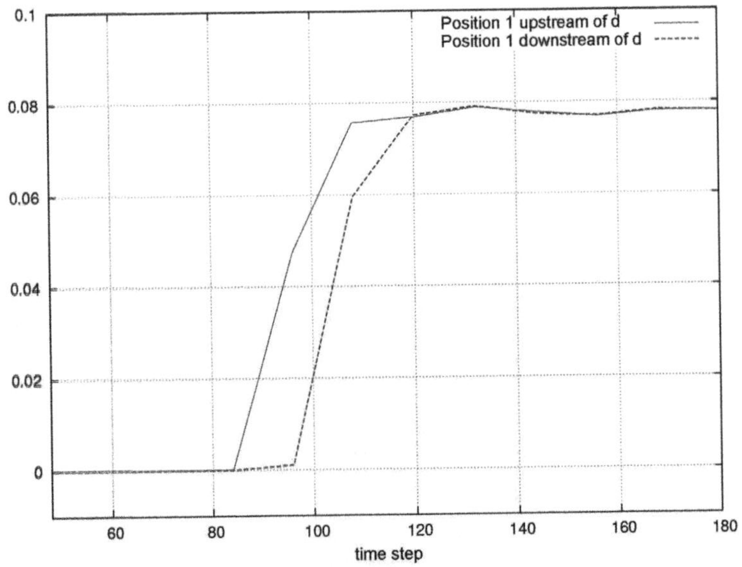
Position 1 upstream of d
Position 1 downstream of d
0.1
0.08
0.06
0.04
0.02
0
60
80
100
120
140
160
180
time step

(c)

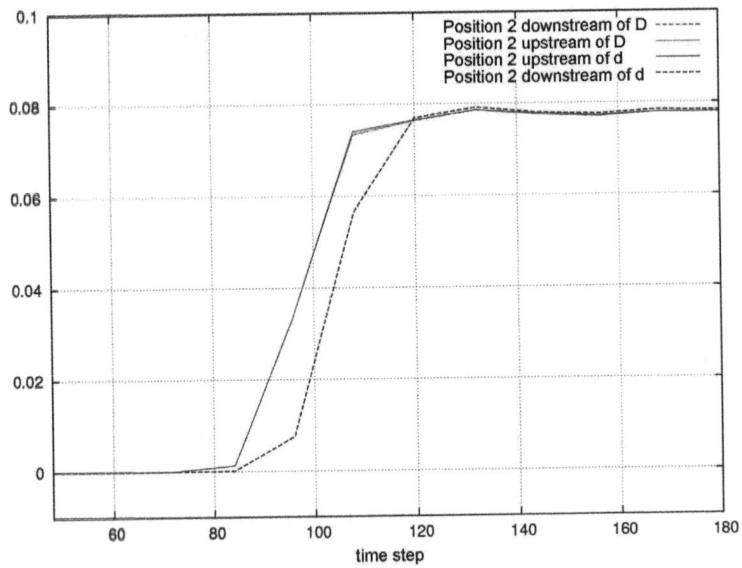
Position 2 downstream of D
Position 2 upstream of D
Position 2 upstream of d
Position 2 downstream of d
0.1
0.08
0.06
0.04
0.02
0
60
80
100
120
140
160
180
time step

(d)

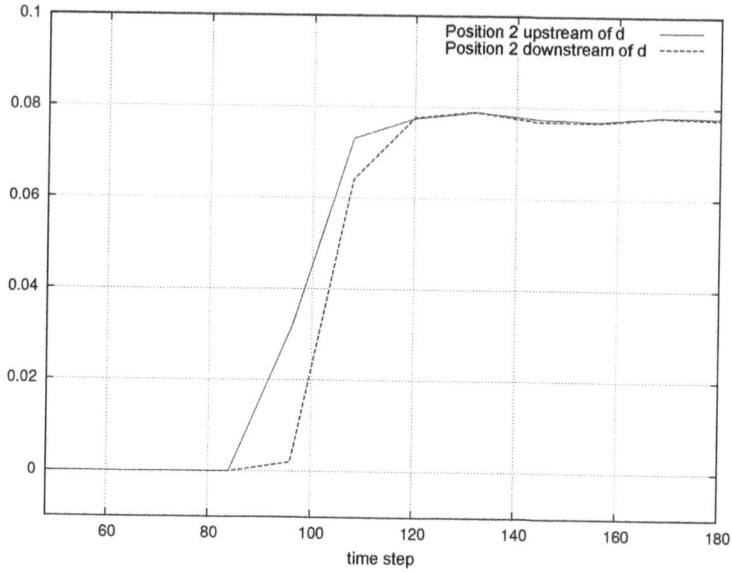

(e)

Figure 3.10. *(a) Mean flow rate for position 1 during the outward movement; (b) mean flow rate for position 1 during the outward movement for a relaxed mesh; (c) mean flow rate for position 1 during the outward movement for a refined mesh; (d) mean flow rate for position 2 during the outward movement for a relaxed mesh; (e) mean flow rate for position 2 during the outward movement for a refined mesh. For a color version of this figure, see www.iste.co.uk/prudhomme/fluid2.zip*

3.3.1.2. *Variation in mean mass*

We wish to calculate the mass $m(V_i, t)$ compressed in each volume V_i for every position i upstream and downstream lying between 1 and 3. The target here is to know whether the dephasage seen on the flow rate profile is due to a physical problem or something else. This mass $m(V_i, t)$ is defined, respectively, over each volume V_i by:

$$\forall i \in \{1, 2, 3\}, m(V_i, t) = \int_{V_i} \rho. \qquad [3.19]$$

Insisting that this mass $m(V_i, t)$ is computed using the integration operator, Figure 3.12 shows the variation in this mass $m(V_i, t)$ over time. This numerical simulation demonstrates that the mass, $m(V_1, t)$, compressed in the volume V_1 is greater than the masses $m(V_2, t)$ and $m(V_3, t)$. This mass $m(V_1, t)$ is perfectly conserved during an established regime. However, it is not conserved during a non-stationary regime. This is not consistent with the dephasage mean flow rate during this period.

Figure 3.11. *Left: volume contained between the upstream position 1 and downstream position 1. Middle: volume contained between the upstream position 2 and downstream position 2. Right: volume contained between the upstream position 3 and downstream position 3. For a color version of this figure, see www.iste.co.uk/prudhomme/fluid2.zip*

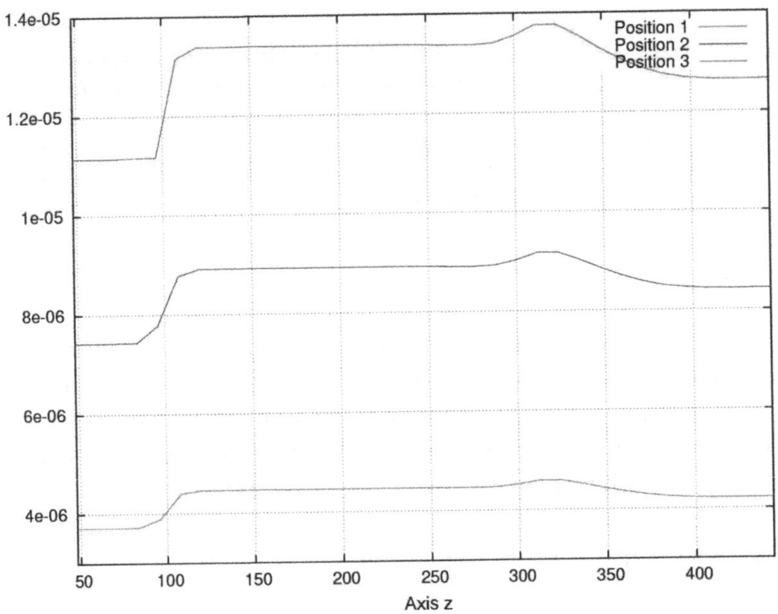

Figure 3.12. *Variation of the mass $m(V_i, t)$ over time. For a color version of this figure, see www.iste.co.uk/prudhomme/fluid2.zip*

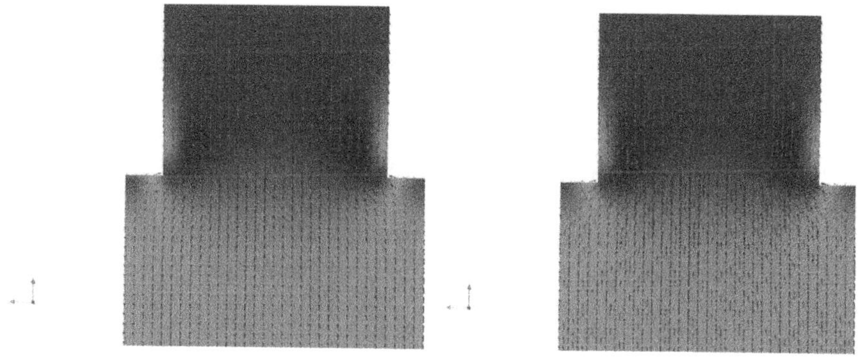

Figure 3.13. *(a) Velocity vector field for a relaxed mesh; (b) velocity vector field for a refined mesh. For a color version of this figure, see www.iste.co.uk/prudhomme/fluid2.zip*

3.3.1.3. *Glyphs and contour line*

This section will explain what happens in the region where there is a change in section. This region is the cause of the non-stationary state of the mean flow rate before the regime is established. Indeed, the goal is to determine the nature of the flow close to the change in section.

The velocity vector field in this region is characterized by its redirection to pass into the second section. The vectors skirt this zone in order to avoid bumping into the edge. With the initial stationary conditions of the Euler equations, circulation zones are not created. However, with non-stationary initial conditions, they are created. For a refined mesh, a randomly distributed vector field is obtained. Further, for a relaxed mesh, the velocity vectors are totally regular. In effect:

– the numerical simulation over a refined mesh shows that the flow is almost 1D and that the velocity field follows a single direction, except along the wall where the flow represents a deviation in velocity along the direction of flow. The velocity field circumvents this section as it passes, retaining the same direction. In the central domain, there is a change in direction for certain velocity vectors, due to 2D flow;

– the direction of flow remains regular, overall, along the axis $\vec{e}_z$ for a relaxed mesh. It is concluded that the flow for a relaxed mesh differs from the flow for a refined one.

Contour lines define the lines as the fluid circumvents them; the horizontal lines show that the flow is ID. However, the encircled lines mean that the flow is in three dimensions in space. For the refined mesh, this represents, relatively, a circle in the second part of the flow. This shows that a flow develops in this region, compared to the same one for a relaxed mesh where there are fewer contour lines.

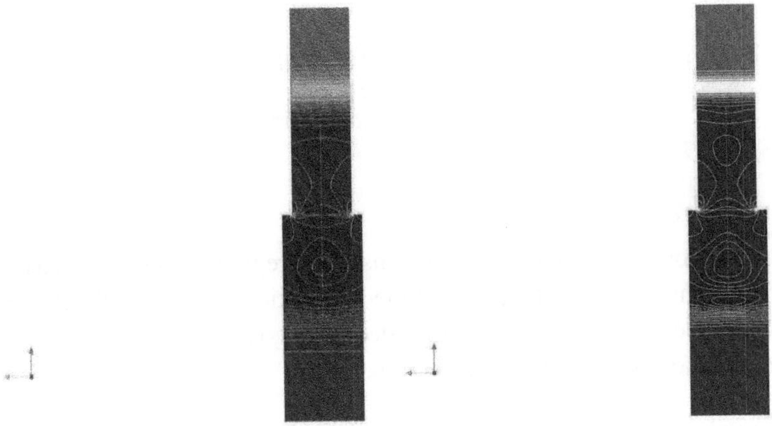

(a) (b)

Figure 3.14. *(a) Contour lines for a relaxed mesh; (b) contour lines for a refined mesh. For a color version of this figure, see www.iste.co.uk/prudhomme/fluid2.zip*

3.3.1.4. *Isosurfaces*

Isosurfaces are used to determine the nature of the 3D flow. As it is difficult to observe this in 2D, we carry out this kind of simulation in order to better observe them across contour lines and the velocity profile. These isosurfaces defined in Figures 3.15a and 3.15b are constructed for greater precision in approximating the space through which the fluid travels. Note that space is divided into several units of volume.

The 3D isosurfaces associated with a relaxed mesh show that the flow is one-dimensional because of the parallel places distributed across the two different sections. However, the 3D contours for a refined mesh confirm that we are in a 2D flow as the planes undergo deformation and cannot, ultimately, be parallel.

(a) (b)

Figure 3.15. *(a) Isosurfaces for a relaxed mesh; (b) isosurfaces for a refined mesh. For a color version of this figure, see www.iste.co.uk/prudhomme/fluid2.zip*

3.3.2. *Results with the boundary and initial conditions II*

3.3.2.1. *Mesh for a quarter of the domain*

Carrying out calculations across the entire domain takes a considerable amount of time and memory. Thus, for the conditions II, it is considered that the domain is quartered. Figure 3.16 computes the duration time in minutes for the different types of meshing across the whole and a quarter of the domain. For a whole domain, Figure 3.17 then shows the time ratio defined by the quotient associated with the computation cost between a coarse mesh and a non-coarse mesh. This ratio is found to be close to 16. On the other hand, for a quartered domain, this ratio is roughly 4.

A relaxed mesh and a refined Cartesian mesh are created across a quarter of the domain to calculate the mean flow rate over a section. This section is defined by the orthogonal one to the flow. This flow rate is then multiplied by four to evaluate the final flow rate. These two meshes according to the time step are defined by:

$$\begin{cases} Relaxed\ mesh, \\ \Delta x = \Delta y = \Delta z = 0.00125m, \\ n_x = 12, n_y = 12, n_z = 80, \\ m_x = 16, m_y = 16, m_z = 80, \\ \Delta t = 1.128 \times 10^{-6}. \end{cases} \quad [3.20]$$

$$\begin{cases} Refined\ mesh, \\ \Delta x = \Delta y = \Delta z = 0.000625m, \\ n_x = 24, n_y = 24, n_z = 160, \\ m_x = 32, m_y = 32, m_z = 160, \\ \Delta t = 0.564 \times 10^{-6}. \end{cases} \quad [3.21]$$

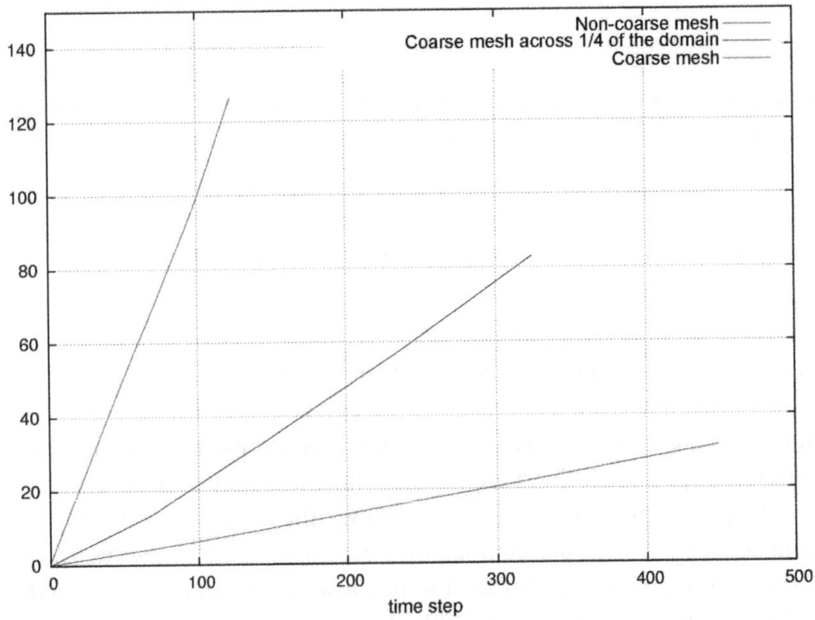

Figure 3.16. *Calculation cost in minutes. For a color version of this figure, see www.iste.co.uk/prudhomme/fluid2.zip*

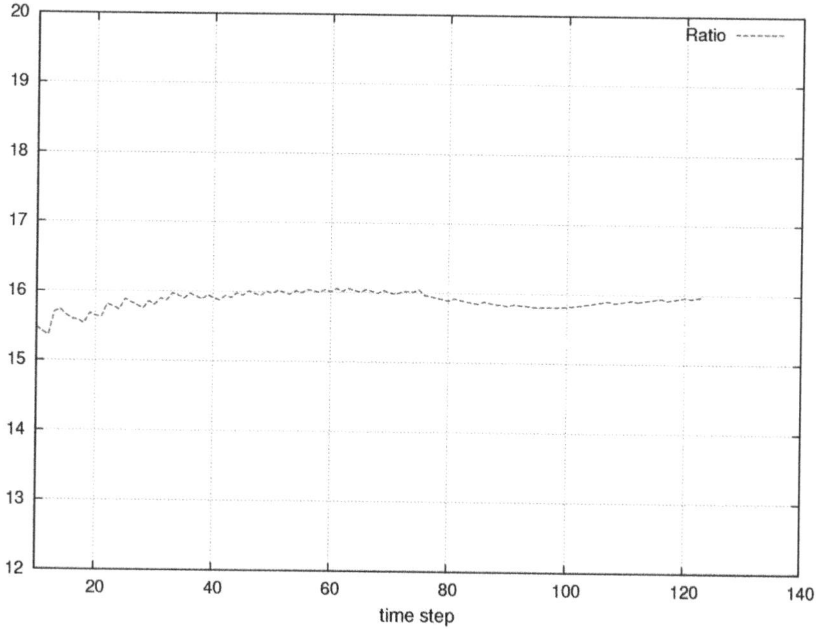

Figure 3.17. *Calculation cost ratio. For a color version of this figure, see www.iste.co.uk/prudhomme/fluid2.zip*

3.3.2.2. *Variations in the mean flow rate*

It has been shown using conditions I that our mesh is indeed convergent, i.e. we obtain the same mean quantities as for the mean flow rate. However, with the initial and boundary conditions, we cannot *a priori* predict that we will find a convergent mesh. Hence, we carry out simulations over two different types of mesh. When compared to conditions I, it can be considered that with conditions II, there is a large difference in temperature and in input velocity.

We will now discuss how the mean flow rate varies over a Cartesian domain with conditions II for the onward and return wave. It can be seen that the mean flow rate largely exceeds 0.3 in the simulation defined in Figure 3.18c for the refined mesh and it does not exceed 0.3 in the simulation defined in Figure 3.18a for a relaxed mesh. On its return crossing, the wave exceeds the value 0.4 for a refined mesh (see Figure 3.18d). However, it does not exceed this value for a relaxed mesh (see Figure 3.18b). Further, the mean flow rate becomes dephasage for a short, non-stationary period. The permanent regime is then established until the end of the return journey.

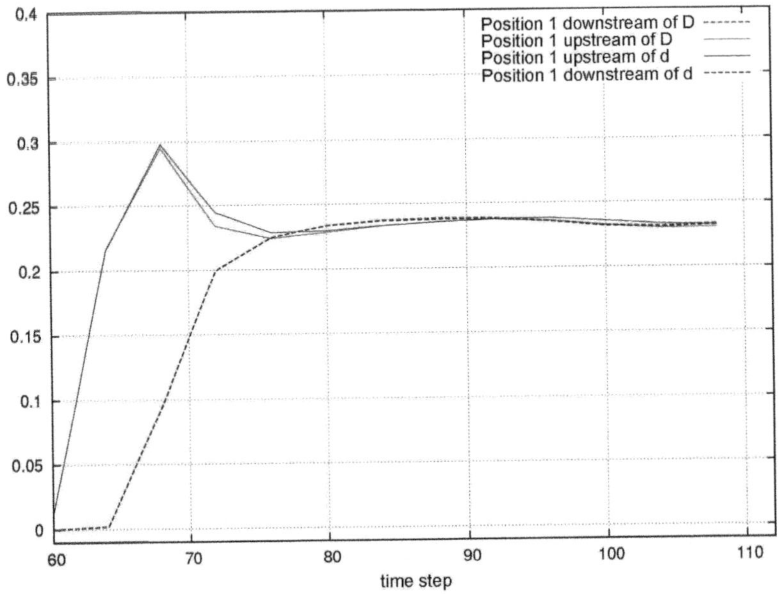
Position 1 downstream of D
Position 1 upstream of D
Position 1 upstream of d
Position 1 downstream of d
0.4
0.35
0.3
0.25
0.2
0.15
0.1
0.05
0
60
70
80
90
100
110
time step

(a)

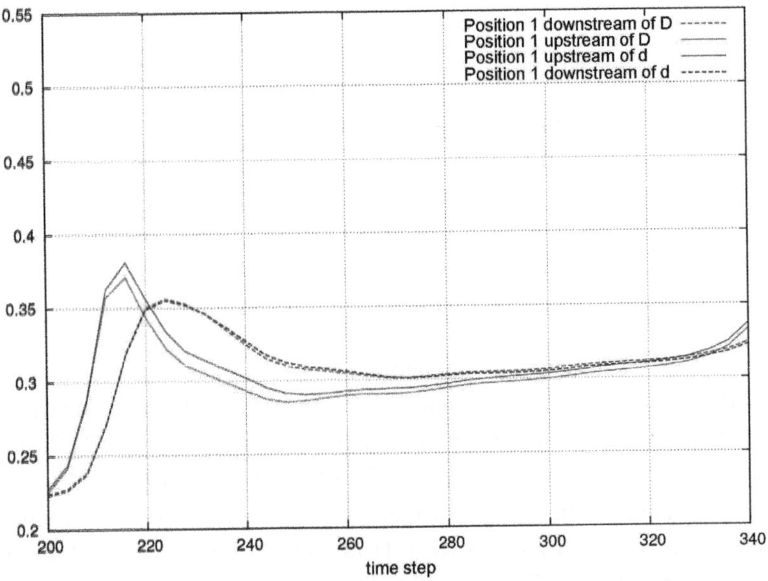
Position 1 downstream of D
Position 1 upstream of D
Position 1 upstream of d
Position 1 downstream of d
0.55
0.5
0.45
0.4
0.35
0.3
0.25
0.2
200
220
240
260
280
300
320
340
time step

(b)

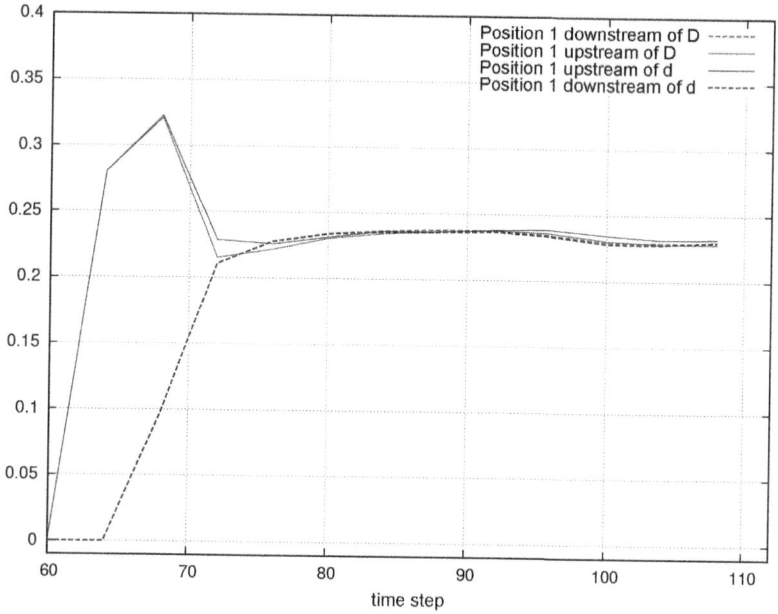
Position 1 downstream of D
Position 1 upstream of D
Position 1 upstream of d
Position 1 downstream of d
0.4
0.35
0.3
0.25
0.2
0.15
0.1
0.05
0
60
70
80
90
100
110
time step

(c)

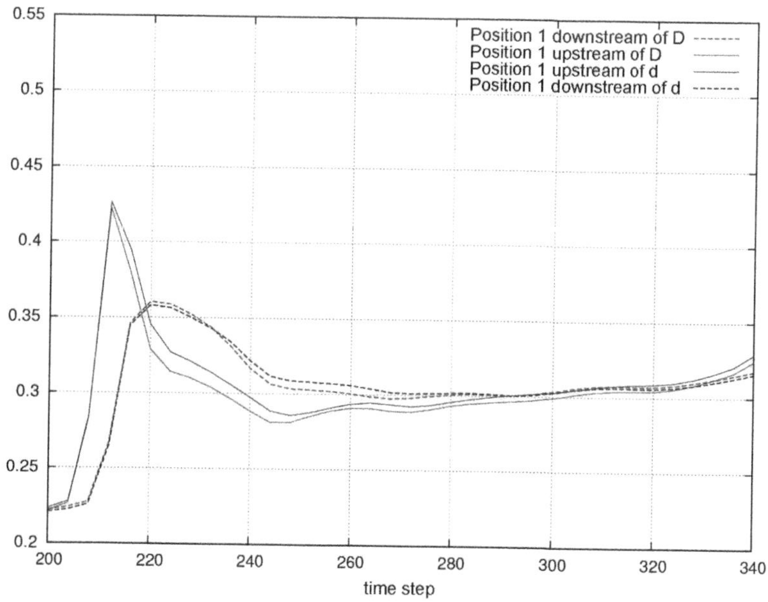
Position 1 downstream of D
Position 1 upstream of D
Position 1 upstream of d
Position 1 downstream of d
0.55
0.5
0.45
0.4
0.35
0.3
0.25
0.2
200
220
240
260
280
300
320
340
time step

(d)

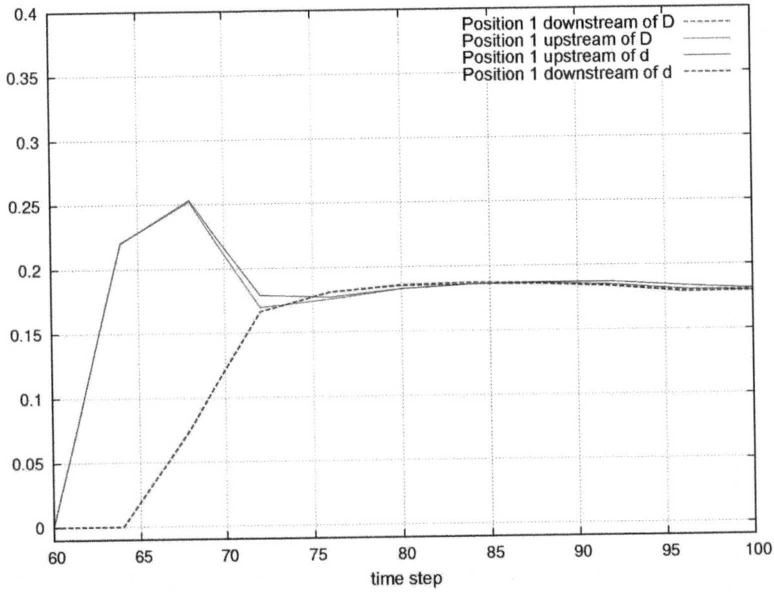

(e)

Figure 3.18. *(a) Mean flow rate for position 1 during an outward movement associated with a relaxed mesh in the Cartesian domain; (b) mean flow rate for position 1 during the return associated with a relaxed mesh in the Cartesian domain; (c) mean flow rate for position 1 during an outward movement associated with a refined mesh in the Cartesian domain; (d) mean flow rate for position 1 during the return associated with a relaxed mesh in the Cartesian domain; (e) mean flow rate for position 1 during an outward movement associated with a refined mesh in a cylindrical domain. For a color version of this figure, see www.iste.co.uk/prudhomme/fluid2.zip*

It is concluded that the structure of the mesh under the conditions II has a considerable impact on the numerical simulation regarding the behavior of the mean flow rate. This happens differently to the conditions I, where the two types of mesh lead to the same numerical results. Furthermore, it is considered that the domain is cylindrical in the simulation defined in Figure 3.18e. We note that for a coarse mesh, the mean flow rate has a maximum value of 0.25. The dephasage remains relatively the same.

3.3.2.3. *The vector field and isosurfaces*

We will now determine the nature of the flow associated with the conditions II. The velocity vector field induces a two-dimensional flow close to the change in section, especially in the second part of the section. As regards a coarse mesh, the velocity vectors move in random directions; more precisely, this happens to the left in the second part, close to the wall. The velocity field circumvents the region close to the wall. The velocity vectors have a regular direction for a non-refined mesh.

We trace isosurfaces for both these types of meshes in Figures 3.19c and 3.19d. For the non-coarse mesh, the simulation defined in Figure 3.19c shows that it is a 1D flow because of the parallel planes. However, for the refined mesh, the simulation defined in Figure 3.19d characterizes a 3D flow, as the planes are not parallel.

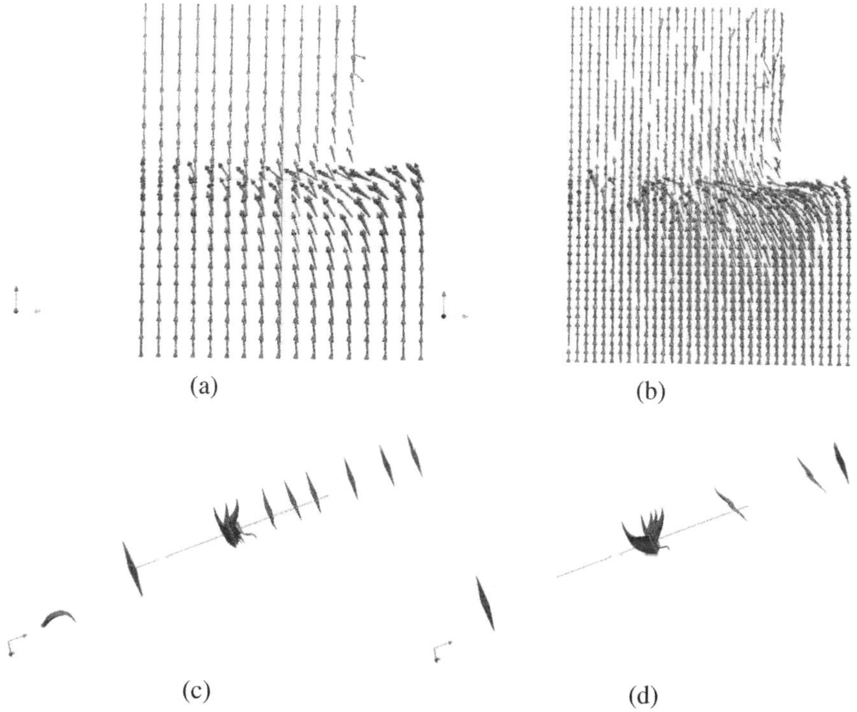

Figure 3.19. *(a) Vector field for a non-refined mesh; (b) vector field for a refined mesh; (c) isosurfaces for a non-refined mesh; (d) isosurfaces for a refined mesh. For a color version of this figure, see www.iste.co.uk/prudhomme/fluid2.zip*

3.4. Conclusion and future prospects

This chapter has demonstrated the need to study models of types of 3D orifices using OpenFOAM, which makes it possible to simulate the flow of gas in evacuation collectors. In future, the objective would be to use a 0D/1D modeling approach on AMESim in order to couple this model and the developed one on OpenFOAM.

3.5. References

Chakkour, T. (2017). Some notes about the continuous-in-time financial model. *Abstract and Applied Analysis* [Online]. Available at: https://doi.org/10.1155/2017/6985820.

Chakkour, T. (2019). Inverse problem stability of a continuous-in-time financial model. *Acta Mathematica Scientia*, 39(5), 1423–1439.

Chakkour, T. and Benkhaldoun, F. (2020). Slurry pipeline for fluid transients in pressurized conduits. *International Journal of Mechanics*, 14 [Online]. Available at: https://www.naun.org/main/NAUN/mechanics/2020/a022003-aci.pdf and https://www.naun.org/cms.action?id=23199.

Dainese, M.P. (1994). Simulation d'écoulements de fluide compressible en géométrie complexe : contribution à l'étude des schémas de discrétisation et d'algorithmes semi-implicites. PhD thesis, ENSAE, Toulouse, France.

Emmert, T., Lafon, P., and Bailly, C. (2009). Numerical study of self-induced transonic flow oscillations behind a sudden duct enlargement. *Physics of Fluids*, 21(10), 106105.

Ersoy, M. (2013). A pressurized model for compressible pipe flow [Online]. Available at: https://hal.archives-ouvertes.fr/file/index/docid/908965/filename/P_Model.pdf.

Guo, B., Fletcher, D.F., and Langrish, T.A.G. (2001). Effect of a downstream sudden contraction on flow instability behind a sudden pipe expansion. *14th Australasian Fluid Mechanics Conference*, Adelaide, Australia, 10–14 December.

Kurganov, A. and Petrova, G. (2000). Central schemes and contact discontinuities. *ESAIM: Mathematical Modelling and Numerical Analysis*, 34(6), 1259–1275.

Kurganov, A. and Tadmor, E. (2000). New high-resolution central schemes for nonlinear conservation laws and convection–diffusion equations. *Journal of Computational Physics*, 160(1), 241–282.

Lacome, J.M., Lemofack, C., Reveillon, J., and Demoulin, F.X. (2012). Numerical modelling of flashing liquid jets due to leakage of liquefied gas storage. *ESReDA Seminar on Land Use Planning and Risk-informed Decision Making*, Rouen, France.

Mujumdar, A.S., Huang, L.X., and Dong Chen, X. (2010). An overview of the recent advances in spray-drying. *Dairy Science & Technology*, 90(2–3), 211–224.

Nessyahu, H. and Tadmor, E. (1990). Non-oscillatory central differencing for hyperbolic conservation laws. *Journal of Computational Physics*, 87(2), 408–463.

OpenFOAM (2016). Open-source Field Operation And Manipulation [Software]. Available at: http://www.openfoam.com [Accessed 26 August 2016].

Ray, S., Biswas, N., and Roy, P.C. (2012). Investigation of Newtonian fluid flow through a two-dimensional sudden expansion and sudden contraction flow passage. *International Journal of Engineering*, 1(12), 01–09.

Satish, G., Ashok Kumar, K., Vara Prasad, V., and Pasha, Sk.M. (2013). Comparison of flow analysis of a sudden and gradual change of pipe diameter using fluent software. *International Journal of Research in Engineering and Technology*, 2(12), 41–45.

4

Vaporization of an Equivalent Pastille

The frequency response of a vaporizing pastille-shaped droplet subjected to small standing acoustic oscillations is examined. The motionless droplet is continuously fed so that its volume remains constant during the process. The feeding can be carried out with the same liquid at different heat exchange coefficients h, and the effects of both heat convection and conduction within the droplet are considered. The effects of variation in the coefficient h value on the evaporating mass response factor and the liquid thermal field are studied.

List of symbols

Symbol	Meaning	Symbol	Meaning
A, B	thermodynamic coefficients	r	radius
c	combustion chamber	R	parameter
$f, \overline{f}$	function, its mean value	$\overline{r}_s$	mean radius of the droplet
$\Delta f, f'$	absolute, relative perturbation $\Delta f/\overline{f}$	s	characteristic eigenvalue
$\hat{f}$	$f' = \hat{f}\exp(i\omega t)$	S	evaporation surface; base area
$G(\xi)$	distribution function	t	time; parameter

Chapter written by Roger PRUD'HOMME and Kwassi ANANI.

Symbol	Meaning	Symbol	Meaning
H, h	thermal exchange coefficient	T	temperature
k	thermal conductivity	u	reduced acoustic pulsation
L, l	indices referring to the liquid phase	v	velocity
ℓ	latent heat of evaporation	V	volume
m, n	exponents in the function G	x	axial coordinate
M	mass of the droplet	Z	transfer function
$\dot{M}$	instantaneous rate of evaporation	α, μ	thermodynamic coefficients
$\overline{\dot{M}}$	mean rate of evaporation	ε	1 with convection; 0 without convection
N	response factor	κ	thermal diffusivity
P	pressure	θ	ratio $\bar{\tau}_v / \tilde{\tau}_T$ (inverse of $\mathrm{Pe_L}$)
Pe	Péclet number	τ	reduced time $\tau = t / \tilde{\tau}_T$
p, q	physical parameters	$\tilde{\tau}_T$	thermal characteristic time
Q	instantaneous heat flux	$\bar{\tau}_v$	mean lifetime (or residence time)
$\Delta \mathrm{Q_L}$	perturbation of Q_L	$\omega, \bar{\tau}_\omega$	pulsation, acoustic period
		ξ	reduced axial coordinate $\xi = x / \bar{x}_S$

4.1. Introduction

We study the dynamic response of a vaporizing pastille-shaped droplet (a small cylinder of liquid called "pastille", Figure 4.1) under the effect of small acoustic oscillations. In such studies, inspired by the Heidmann (1972) model for spherical droplets (Prud'homme *et al.* 2010), the pastille is fed continuously – adiabatically or isothermally – and we take into account the effects of heat convection and conduction on the system's frequency response factor and the thermal field in the droplet. In the present work, the pastille is fed and has a thermal exchange coefficient, and this feeding is also no longer restricted to boundary adiabatic or isothermal cases.

The equivalent pastille was introduced in a simplified model (Anani *et al.* 2011) of HF (high-frequency) instabilities in the combustion of droplets. Indeed, the solution to a non-stationary problem with spherical symmetry between the variables r and t must be replaced by a 1D non-stationary problem with the variables x and t.

The advantage here is that equations for small perturbations in the pastille subject to an acoustic field are linear, with constant coefficients, and thus present analytical solutions.

Compared to a spherical droplet with mean diameter $2\overline{r}_s$, an equivalent pastille has the same volume and only vaporizes over its basic surfaces, as shown in Figure 4.1.

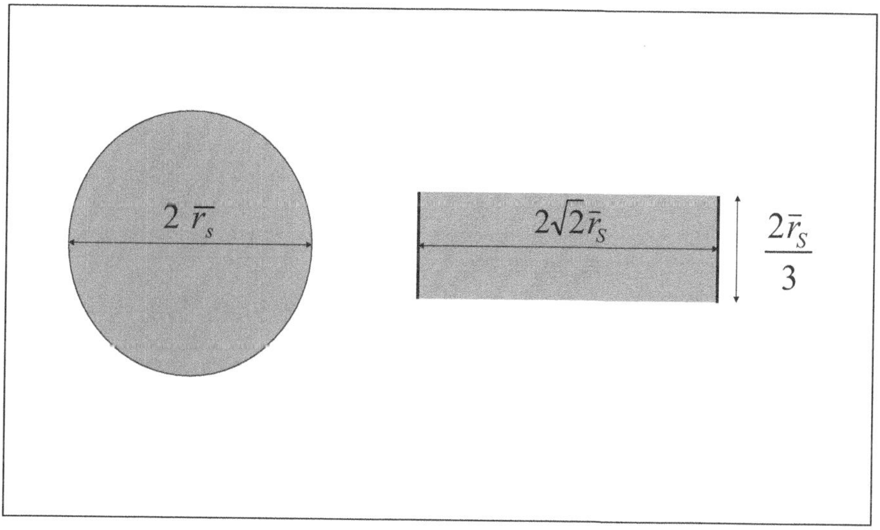

Figure 4.1. *Geometric configuration of a pastille equivalent to a sphere with diameter $2\overline{r}_s$. The lateral wall of the cylinder is rigid, impermeable and adiabatic, while the spherical droplet exchanges mass and heat across its whole surface*

To facilitate the implementation of the equations, we choose to consider the equivalent pastille shown in Figure 4.2 (adapted from Anani *et al.* (2011)).

We will focus on the conditions under which the pastille-shaped droplet is fed. The injected mass flow rate is always the mean rate of evaporation $\bar{\dot{M}}$; however, the thermal conditions may be different. We characterize them with a single exchange coefficient, H. Only the adiabatic injection ($H = 0$) and the isothermal injection ($H = \infty$) have been studied up until now.

For the gaseous phase, we will propose the same hypotheses as those in the earlier work: a mixture made up of the vapor of the droplet F and the burnt gas A, knowing that the complete combustion takes place far from the droplet in the combustion chamber (index c), where the quantities f have the local value $\bar{f}$, a state perturbed in a harmonic manner by a standing acoustic wave, such that: $f = \bar{f} + \Delta f$, $f' = \Delta f / \bar{f}$, $|\Delta f / \bar{f}| << 1$, and $f'(\mathbf{x},t) = \hat{f}(\mathbf{x}) e^{i\omega t}$. It is assumed that the droplet is positioned at a pressure crest (which is also a velocity node) with a maximal amplitude $\hat{p}_c$. It can be demonstrated that at the level of the surface of the droplet, its evolution is governed by a system of two equations:

$$\hat{\dot{M}} = \alpha \frac{iu}{1+iu}\left(\bar{b}\hat{T}_S - \hat{p}_c\right),\ \Delta\hat{Q}_L = \bar{\dot{M}}\,\bar{\ell}\left(\bar{a}\,\hat{p}_c - \mu \hat{T}_S\right) \qquad [4.1]$$

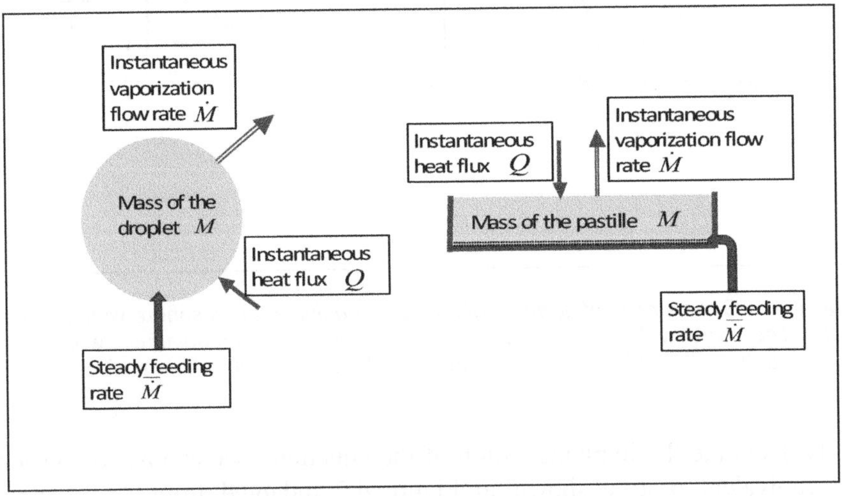

Figure 4.2. *Droplet with radius $\bar{r}_s$ fed with liquid with a flow rate $\bar{\dot{M}}$ and the equivalent pastille. The liquid is fed in from the bottom of the liquid pastille, through a porous, cylindrical plate with the same diameter as the pastille. For a color version of this figure, see www.iste.co.uk/prudhomme/fluid2.zip*

In system [4.1], $\dot{M}, Q_L, T_S, p_c, \ell$ are, respectively, the mass flow rate of evaporation, the heat flow penetrating into the droplet (which was not used in evaporation), the surface temperature, the pressure of the chamber and the latent heat of evaporation. The coefficients involved in these equations depend on the values of the thermodynamic and chemical reference quantities of the undisturbed mixture in the chamber and at the level of the evaporation surface. The expressions for these coefficients can be found in the articles mentioned in Appendix A4.3 in the book by Prud'homme (2012). The equations for system [4.1] will be coupled with the equation for the liquid phase in the final solution.

4.2. Equations for the problem

The liquid phase is assumed to be incompressible and non-dilatable. Its thermodynamic constants are assumed to be constant.

The ***notations*** are as follows:

– the mean mass for the cylindrical droplet $\overline{M} = S\rho_L \bar{x}_S$ (with $\bar{x}_S$ being the mean thickness);

– the mean mass rate flow for the supply (or of evaporation) $\overline{\dot{M}} = S\rho_L v_L$;

– the mean residence time (or lifetime) of the liquid $\bar{\tau}_v = \overline{M}/\overline{\dot{M}} = \bar{x}_S/v_L$;

– the characteristic thermal conduction time $\tilde{\tau}_T = \bar{x}_S{}^2/\kappa_L$ (with $\kappa_L = k_L/\rho_L c_L$);

– the Péclet number in the liquid $\text{Pe}_L = \tilde{\tau}_T/\bar{\tau}_v$ (its inverse θ is used in certain calculations).

The ***basic equations*** of the liquid phase are:

– the mass balance equation for the pastille: $dM/dt = \overline{\dot{M}} - \dot{M}$;

– the local balance equation for the internal energy: $\partial T/\partial t + v_L \partial T/\partial x - \kappa_L \partial^2 T/\partial x^2 = 0$.

The ***boundary conditions*** are written as:

– at $x = x_S$ (free surface), $T_l(x_S,t)=\overline{T}_S(t)$, $S\,k_L\,\partial T_l/\partial x|_{\overline{x}_S}=Q_L$;

– at x = 0 (at the bottom of the cylinder where the supply comes in), $\partial T_l/\partial x|_0 = H(T_l-\overline{T}_S)$. For an adiabatic supply, H = 0 and if the supply is isothermal, $H = \infty$.

4.3. Linear analysis of the liquid phase

4.3.1. *The function* $G(u, Pe_L)$

As was indicated in the introduction, we write: $f=\bar{f}+\Delta f$ and $f'=\Delta f/\bar{f}$, $|\Delta f/\bar{f}|<<1$.

The equation for energy becomes: $\partial T'/\partial t + v_L\,\partial T'/\partial x - \kappa_L\,\partial^2 T'/\partial x^2 = 0$ and the condition at the free surface can be written as: $\pi\,\overline{r}_c^{\,2}\,k_L\,\partial T_l'/\partial x|_{\overline{x}_S} = \Delta Q_L$,

with $\overline{r}_c = 2\overline{r}_S$.

By writing $f'(x,t)=\hat{f}(x)e^{i\omega t}$, we obtain: $i\,\omega + \dfrac{s\,x_S}{\tau_v} - \dfrac{(s\,x_S)^2}{\tilde{\tau}_T} = 0$, or again:

$$(s\,x_S)^2 - \varepsilon\,Pe_L\,(s\,x_S) - i\,Pe_L u/3 = 0$$

with $u = 3\omega\tilde{\tau}_T$, $\varepsilon = 1$ when convection is present and $\varepsilon = O$ when neglecting convection. This characteristic equation has two roots s^+ and s^-. We thus express the perturbation in temperature using the quantities: $m = s^+\overline{x}_S$, $n = s^-\overline{x}_S$, $\xi = x/\overline{x}_S$, $h = \overline{x}_S\,H$:

$$\hat{T}_l(\xi) = \hat{T}_S\,\frac{e^{m\xi}/(m+h) - e^{n\xi}/(n+h)}{e^{m}/(m+h) - e^{n}/(n+h)} \qquad [4.2]$$

The condition at the free surface can be written as $\Delta\hat{Q}_L = S\,k_L\,\overline{T}_S\,\partial\hat{T}_l/\partial x|_{\overline{x}_S} = (S\,k_L\,\overline{T}\,\overline{T}_S/\overline{x}_S)G(u,Pe_L)$, and we have:

$$G(u,Pe_L)=\frac{\frac{m}{m+h}e^m-\frac{n}{n+h}e^n}{\frac{1}{m+h}e^m-\frac{1}{n+h}e^n} \qquad [4.3]$$

4.3.2. *Solution*

The coefficients m and n are connected to the quantities ε, u, θ through the following sequence:

$$\begin{aligned}&t=-\frac{3\varepsilon\,\mathrm{Pe}_L}{4\,u}+\sqrt{1+\left(\frac{3\varepsilon\,\mathrm{Pe}_L}{4\,u}\right)^2},\quad R=\sqrt{\left(\frac{\varepsilon\,\mathrm{Pe}_L{}^2}{4}\right)^2+\left(\frac{u\,\mathrm{Pe}_L}{3}\right)^2}\\&m=\frac{\varepsilon\,\mathrm{Pe}_L}{2}+\sqrt{R}\frac{1+it}{\sqrt{1+t^2}},\quad n=\frac{\varepsilon\mathrm{Pe}_L}{2}-\sqrt{R}\frac{1+it}{\sqrt{1+t^2}}\end{aligned} \qquad [4.4]$$

We derive:

$\hat{T}_S=\frac{1}{b}\left(1+\frac{1+iu}{iu}Z\right)\hat{p}_c$, with the complex transfer function Z:

$$Z=\frac{iu}{1+iu}\,\frac{A-3\theta G(u,\mathrm{Pe}_L)}{B+3\theta G(u,\mathrm{Pe}_L)} \qquad [4.5]$$

It must be recalled that the response factor, N, is defined from:

$$N=\frac{\iint_{V,t}q'(V,t)p'(V,t)dt\,dV}{\iint_{V,t}(p'(V,t))^2\,dt\,dV} \qquad [4.6]$$

where p' is the incidental perturbation (here the perturbation in the chamber's pressure $p'c$) and q' is the system's response (here the perturbation in the flow rate of the droplet, $\dot{M}'$) to this perturbation (Rayleigh criterion (Rayleigh 1945)).

The reduced response factor is then equal to the real part of Z:

$$\frac{N}{\alpha} = \Re(Z) \tag{4.7}$$

By defining the perturbation in the temperature of the liquid as being dependent on time through $\hat{T}_{l\,réd} = \frac{\overline{b}}{\hat{p}_c}\hat{T}_l$, we obtain:

$$\hat{T}_{l\,réd} = \frac{A+B}{B+3\theta G}\frac{e^{m\xi}/(m+h)-e^{n\xi}/(n+h)}{e^{m}/(m+h)-e^{n}/(n+h)} \tag{4.8}$$

Moreover, the real thermal perturbation within the pastille is written as:

$$T'_{l\,réd} = \Re\left(e^{\frac{iux}{3}}\hat{T}_{l\,réd}\right) = \Re\left(\frac{A+B}{B+3\theta G}e^{\frac{iux}{3}}\frac{e^{m\xi}/(m+h)-e^{n\xi}/(n+h)}{e^{m}/(m+h)-e^{n}/(n+h)}\right) \tag{4.9}$$

4.3.3. *The depth to which heat penetrates*

Through a first-order expansion with $\Delta\xi$, we obtain[1]:

$$T'_l \propto 1 + \frac{\frac{m}{m+h}e^m - \frac{n}{n+h}e^n}{\frac{1}{m+h}e^m - \frac{1}{n+h}e^n}\Delta\xi\, N/\alpha \tag{4.10}$$

The temperature variation curve intersects the ξ axes when:

$$|\Delta\xi| \approx \left|\frac{\frac{1}{m+h}e^m - \frac{1}{n+h}e^n}{\frac{m}{m+h}e^m - \frac{n}{n+h}e^n}\right| = \left|\frac{1}{G}\right| \tag{4.11}$$

and yields an evaluation of the depth to which the acoustic wave has penetrated into the droplet when this quantity is low.

1 This limited expansion only has values for small penetration depths $\Delta\xi$.

4.4. Some results

	Without thermal convection	With thermal convection
Adiabatic bottom	$\varepsilon = 0,\ h = 0$	$\varepsilon = 1,\ h = 0$
Exchange coefficient	$\varepsilon = 0,\ h > 0$	$\varepsilon = 1,\ h > 0$
Isothermal bottom	$\varepsilon = 0,\ h = \infty$	$\varepsilon = 1,\ h = \infty$

Table 4.1. *Six types of configuration when there is an exchange coefficient at the heart of the supply*

Table 4.1 presents different configurations that are possible with our model.

The results obtained for $h = 0$ and $h = \infty$ were analyzed by Anani (2012). We will now study the influences of the exchange coefficients for non-null finite values on the curves for the reduced response factor N/α and the reduced thermal perturbation profiles $T'_{l\,réd}$ within the pastille.

4.4.1. *Thermal perturbations*

The two parameters that influence heat conductivity are the Péclet number, Pe_L, between the top and the bottom of the pastille and the exchange coefficient h at the bottom of the pastille, at the level where the supply liquid comes in, which has the temperature $\overline{T}_S$. It is indeed verified that Pe_L controls the penetration of the thermal perturbation from the surface, while h has a decisive action at the bottom.

In Figure 4.3, it can be observed that when the Péclet number is constant throughout the body of the liquid, there is a strong influence of the thermal exchange coefficient, h, at the bottom of the pastille for $\xi = 0$. When $h=0$, the bottom is adiabatic, and it can be seen that the thermal profile, over time, observed at the surface of the pastille at $\xi = 1$, is reproduced in its entirety at the bottom since the thermal transfer in the liquid allows this non-change, given the value of the Péclet number. Conversely, the isothermal bottom enforces the temperature of the supply liquid and $T'_{l\,réd} = 0$.

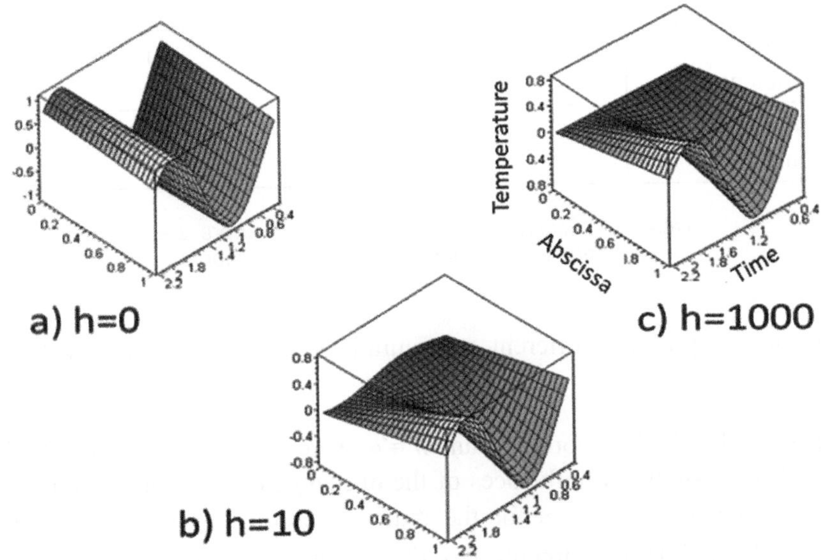

Figure 4.3. *Influence of the exchange coefficient with the values being identical for all the other parameters* $\varepsilon = 1, \mathrm{Pe}_L = 0{,}1(\theta = 10), u = 10\ A = 10, B = 100$. *The axes correspond to the reduced_coordinates: temperature* $T'_{lréd}$ *(equation [4.9]), abscissa* $\xi = x / x_s$ *and time* $\tau = t / \tau_v$. *For a color version of this figure, see www.iste.co.uk/prudhomme/fluid2.zip*

Figure 4.4 shows the influence of the Péclet number $\mathrm{Pe}_L = \widetilde{\tau}_T / \overline{\tau}_v$ of the liquid in the pastille. For a relatively large Pe_L, the temperature of the supply, $\overline{T}_S$, is imposed by convection across 80% of its thickness. For smaller values of Pe_L, the temperature of the surface of the pastille penetrates deeper into the pastille, without reaching the situation in Figure 4.3a, given that the exchange coefficient h is non-zero this time.

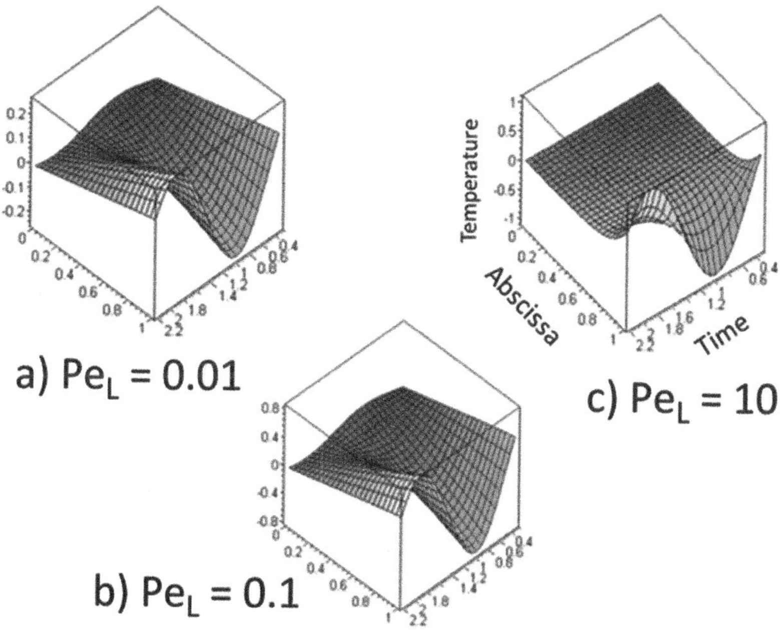

Figure 4.4. *Influence of the Péclet number (inverse of θ) with the values being identical for all the other parameters* $\varepsilon = 1, \mathrm{h} = 10, u = 10, A = 10, B = 100$. *The axes correspond to the reduced coordinates: temperature* $T'_{lréd}$ *(equation [4.9]), abscissa* $\xi = x / x_s$ *and time* $\tau = t / \tau_v$. *For a color version of this figure, see www.iste.co.uk/prudhomme/fluid2.zip*

4.4.2. *Response factor*

The response factor N indicates the stability state of an evolution that is a function of the pulsation ω. If this is positive, it is unstable, and if it is negative, it is stable. This result is valid for reduced quantities N/α and u. Figures 4.5 and 4.6 present the calculation results obtained using MAPLE 8. These results include the values of the exchange coefficient h = 0; 10; 1,000, as well as the values for θ and the Péclet number Pe_L = 0.01; 0.1; 10. explored in Figure 4.4.

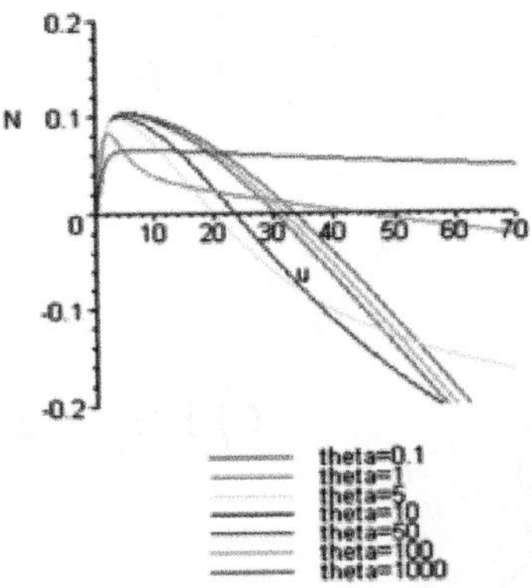

h = 0 (adiabatic bottom)

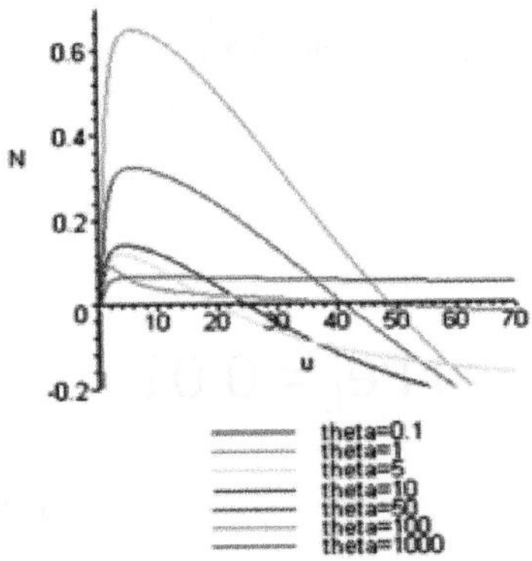

h = 0.1

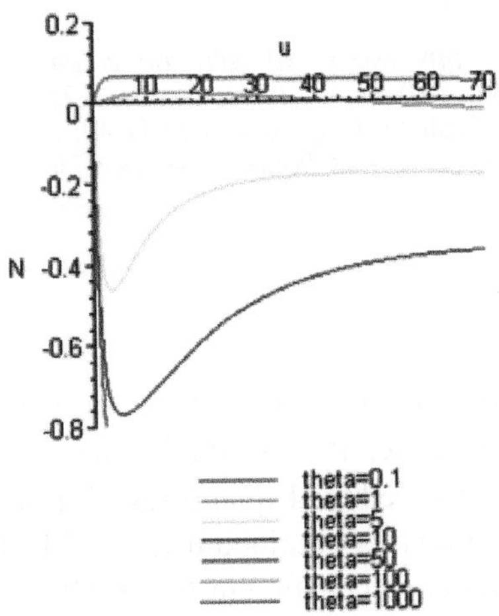

h = 1

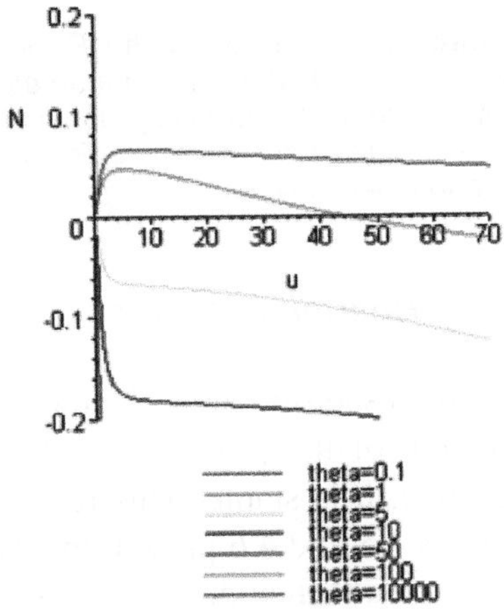

h = 10

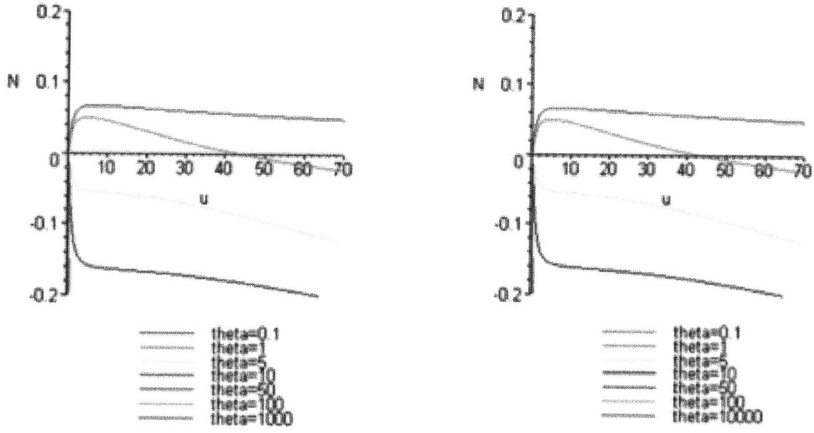

h = 100　　　　　h = 1000 (isothermal bottom)

Figure 4.5. *Curves showing the reduced response factor N/α as a function of the reduced pulsation of the harmonic perturbation of pressure in the combustion chamber, obtained by taking into account the effects of convection ($\varepsilon = 1$). For a color version of this figure, see www.iste.co.uk/prudhomme/fluid2.zip*

If the internal convection of the liquid is considered (as in section 4.1, where we took $\varepsilon = 1$), we obtain Figure 4.5.

For an adiabatic pastille bottom (h = 0), and for increasing values of $\theta = 1/Pe_L$, there is initially a large unstable zone θ = 0.1 whose size (given by the value u_C of the intersection with the axis N = 0) first decreases as a function of θ and then increases until a limiting value for an infinite value of θ is reached. This limiting value corresponds to an infinitely rapid thermal conduction within the liquid. For h = 0.1, the curve θ = 1,000 disappears from the upper half-plane to pass into the region for $N < 0$ (stability). Then, only the curves θ = 0.1 and θ = 1 present positive values (thus the instability zone). Finally, for higher values of h (h = 10 and h = 1,000, which indicate isothermal supply), we observe complete stabilization, notably the stability set in for values of $\theta \geq 5$.

The general interpretation of the results is not simple. Analogous areas between injection regimes with an adiabatic bottom and those with an isothermal bottom have been identified in Anani *et al.* (2011) (see Figure 4.6). It can be seen here that large differences between the two regimes arise only if the thermal perturbation reaches the bottom of the pastille after thermal diffusion has occurred within this pastille.

This is possible if the thermal diffusion time is smaller than an oscillation period and the time of residence $\overline{\tau}_v$. The calculations carried out for the pastille with $A = 10$ and $B = 100$ give $\theta^* = 5$ and $u^* \approx 15$.

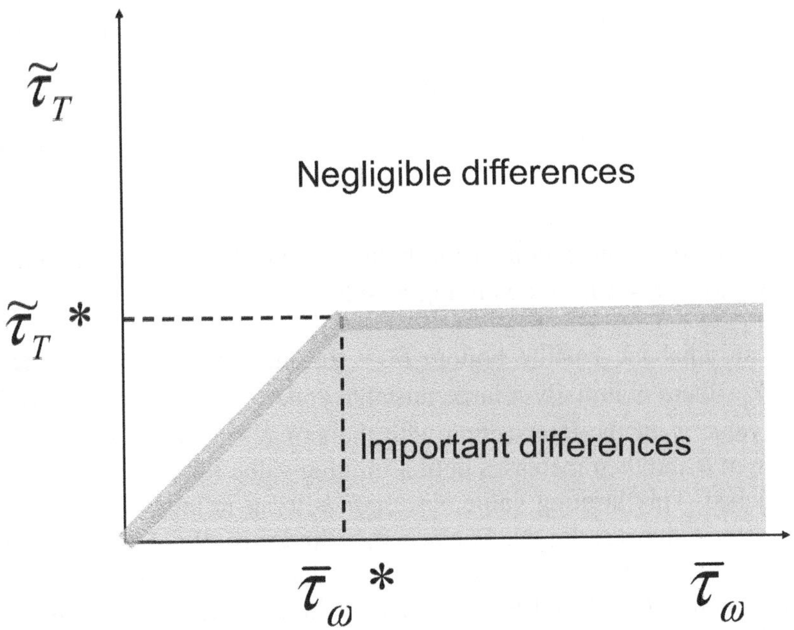

Figure 4.6. *Differences between the temperature field in the liquid for two conditions at $x = 0$: isothermal (enforced temperature) and adiabatic (zero thermal conduction). $\tau_\omega = 2\pi/\omega$ denotes the period of harmonic oscillation (Anani* et al. *2011). For a color version of this figure, see www.iste.co.uk/prudhomme/fluid2.zip*

4.5. Conclusion

This study makes it possible to refine the results with respect to the effects of the liquid supply to a "mean droplet" in an engine, by introducing an exchange coefficient. Indeed, the previous studies only considered the limiting cases with adiabatic and isothermal injection. It can thus be seen that we must take precautions when refining the Heidmann configuration, taking into account the thermal field within the mean droplet that is assumed to represent the suspension in the combustion within a liquid-propellant engine. Experimental verification must, of course, be carried out. However, the introduction of the exchange coefficient brings a new flexibility to the model, bringing it closer to reality.

However, the physical aspects related to this mean droplet may help us to better understand the behavior of the two-phase flow itself. This is the case for the characteristic time $\overline{\tau}_v$, which is both the time of residence of the liquid between the point of supply and the evaporation surface and the mean lifetime of the droplets in suspension in combustion. Bringing in the Péclet number for the liquid phase certainly makes sense for the flow of the suspension as well as the supply parameters of the mean droplet.

What remains is to adapt this model to the case of a spherical evaporation interface.

The calculations presented in this chapter and the resulting curve diagrams were carried out using the Maple programs.

The model for the mean droplet was also used for an irreversible thermodynamic application sketched out by Prud'homme (2012, pp. 282–288).

It was used as a reference to perfect the CEDRE code in ONERA (Prud'homme *et al.* 2010).

Recent works related to this topic have been published (Anani and Prud'homme 2016).

Other more general publications may be useful in complementing the literature on the subject of instabilities in the presence of vaporizing droplets and droplets in combustion (Abramzon and Sirignano 1989; Candel 1992; Sazhin *et al.* 2006; Sirignano 2015).

4.6. References

Abramzon, B. and Sirignano, W.A. (1989). Droplet vaporization model for spray combustion calculations. *International Journal of Heat and Mass Transfer*, 32(9), 1605–1618.

Anani, K. (2012). Contribution à l'étude de l'injection d'un fluide dans une goutte soumise aux instabilités acoustiques de combustion. PhD thesis, University of Lomé.

Anani, K. and Prud'homme, R. (2016). Theoretical analysis of thermal conduction effect on frequency response of a perturbed vaporizing spherical droplet. *Flow, Turbulence and Combustion*, 98(2), 503–522.

Anani, K., Prud'homme, R., D'Almeida, A., Assiamoua, K. (2011). Effect of thermal convection on frequency response of a perturbed vaporizing pastille-shaped droplet. *Mécanique & industries*, 12(4), 301–314.

Candel, S.M. (1992). Combustion instabilities coupled by pressure waves and their active control. *Symposium (International) on Combustion*, 24(1), 1277–1296.

Heidmann, M.F. (1972). Frequency response of a vaporization process to distorted acoustic disturbances. NASA Technical Note, D-6806.

Prud'homme, R. (2012). *Flows and Chemical Reactions*. ISTE Ltd, London and John Wiley & Sons, New York.

Prud'homme, R., Habiballah, M., Matuszewski, L., Mauriot, Y., Nicole, A. (2010). Theoretical analysis of dynamic response of a vaporizing droplet to acoustic oscillations. *Journal of Propulsion and Power*, 26(1), 74–83.

Rayleigh, L. (1945). *Theory of Sound*. Dover Publications, New York.

Sazhin, S.S., Kristyadi, T., Abdelghaffar, W.A., Heikal, M.R. (2006). Models for fuel droplet heating and evaporation: Comparative analysis. *Fuel*, 85(12–13), 1613–1630.

Sirignano, W.A. (2015). Driving mechanisms for combustion instability. *Combustion Science and Technology*, 187(1–2), 162–205.

5

Thermal Field of a Continuously-Fed Drop Subjected to HF Perturbations

The oscillatory temperature field of a spherical drop in evaporation is examined. The motionless drop is fed continuously at the center so that it maintains a constant volume during the process. The feeding with the same liquid is carried out by assuming an adiabatic center regime. The effects of heat conduction and that of the variation of the Péclet number on the thermal field inside the droplet are studied.

List of symbols

Symbol	Meaning	Symbol	Meaning
A, B	thermodynamic coefficients	r	radius; radial coordinate
B_M, B_T	spalding parameters	$\bar{r}_s$	mean radius of the drop
c	combustion chamber	s	characteristic Eigenvalue
$f, \bar{f}$	function, its mean value	S	evaporation surface: base area
$\Delta f, f'$	absolute, relative perturbation $\Delta f / \bar{f}$	t	time

Chapter written by Roger PRUD'HOMME, Kwassi ANANI and Mahouton Norbert HOUNKONNOU.

Symbol	Meaning	Symbol	Meaning
$\hat{f}$	$f' = \hat{f}\ \exp(i\omega t)$	T	temperature
$E(u,\theta)$	distribution function	u	reduced acoustic pulsation
H,h	thermal heat coefficient	v	velocity
k	thermal conductivity liquid phase	V	volume
L,l	indices referring to the liquid phase	x	axial coordinate
ℓ	latent heat of evaporation	Z	transfer function
M	mass of the drop	α, μ	thermodynamic coefficients
$\dot{M}$	instantaneous rate of evaporation	κ	thermal diffusivity
$\overline{\dot{M}}$	mean rate of evaporation	θ	ratio $\bar{\tau}_v / \tilde{\tau}_T$ (inverse of $\mathrm{Pe_L}$)
N	response factor	τ	reduced time $\tau = t / \bar{\tau}_v$
p	pressure	$\tilde{\tau}_T$	characteristic thermal time
Pe	Péclet number	$\bar{\tau}_v$	mean lifetime (or time of residence)
p,q	physical parameters	$\omega, \bar{\tau}_\omega$	acoustic pulsation, period
Q	instantaneous heat flow	ξ	reduced axial coordinate $\xi = x / \bar{x}_S$; reduced radial coordinate $\xi = r / \bar{r}_S$
$\Delta \mathrm{Q_L}$	Perturbation of Q_L		

5.1. Drops in a liquid-propellant rocket engine

Figure 5.1 shows the physical phenomena brought into play in a cryogenic, coaxial injector flame. The liquid oxygen is injected into a gaseous hydrogen jet.

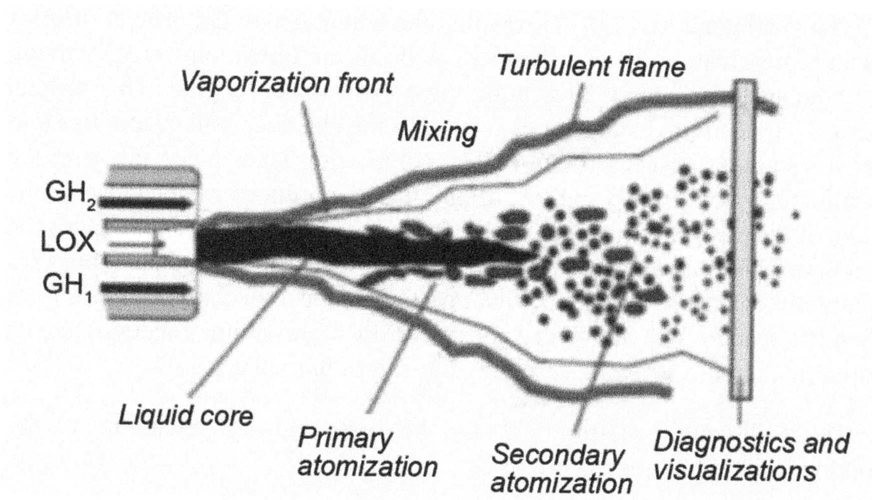

Figure 5.1. *Physical phenomena brought into play in a cryogenic, coaxial injector flame (ONERA diagram). For a color version of this figure, see www.iste.co.uk/prudhomme/fluid2.zip*

This diagrammatic representation shows the complexity of the physical phenomena to be considered if a combustion instability, for example, is to be modeled. This explains why simplified models are used.

It is known that evaporation may amplify HF (high-frequency) acoustic perturbations arising from the rocket engine and damage the rocket. We will therefore start by modeling what happens at the level of the drops themselves when they are subjected to a harmonic acoustic perturbation with a pre-set frequency and amplitude. Instead of a detailed study of the two-phase flow, we consider a single drop that is a mean representative of what happens. This was the motivation behind studying a continuously fed drop.

5.2. A continuously fed droplet

The Heidmann idea (1972) was that the continuously fed droplet offers a mean representation of what happens with all the drops that travel through the combustion chamber of a liquid-propellant rocket engine. This droplet has a fixed position, located at the center of the chamber, and evaporates into the surrounding gaseous mixture. The combustion takes place far from the continuously fed droplet and the flame that is produced rather close to the walls of the chambers (designated by the index C in Figure 5.2) produces the conditions of burned gases T_C and p_C. For the drop, this data represents conditions at infinity. Figure 5.2a shows the combustion chamber as a sort of chemical reactor with an inlet (the green arrow representing injection) and an outlet (the orange arrow representing the exit of burned gases).

The following equation [5.1] is the mass balance equation for the continuously fed droplet:

$$\frac{dM}{dt} = -\dot{M} + \overline{\dot{M}} \quad [5.1]$$

As the density of the liquid is assumed to be constant and uniform, we have

$$M = \frac{4}{3}\pi r_S^3 \rho_L$$

In the following sections, we will describe what happens in the case of a continuously fed droplet. We will first look at what happens in the gaseous phase surrounding the drop, from its surface (index S) to infinity (index C). We will then study the liquid phase, i.e. the interior of the continuously fed drop.

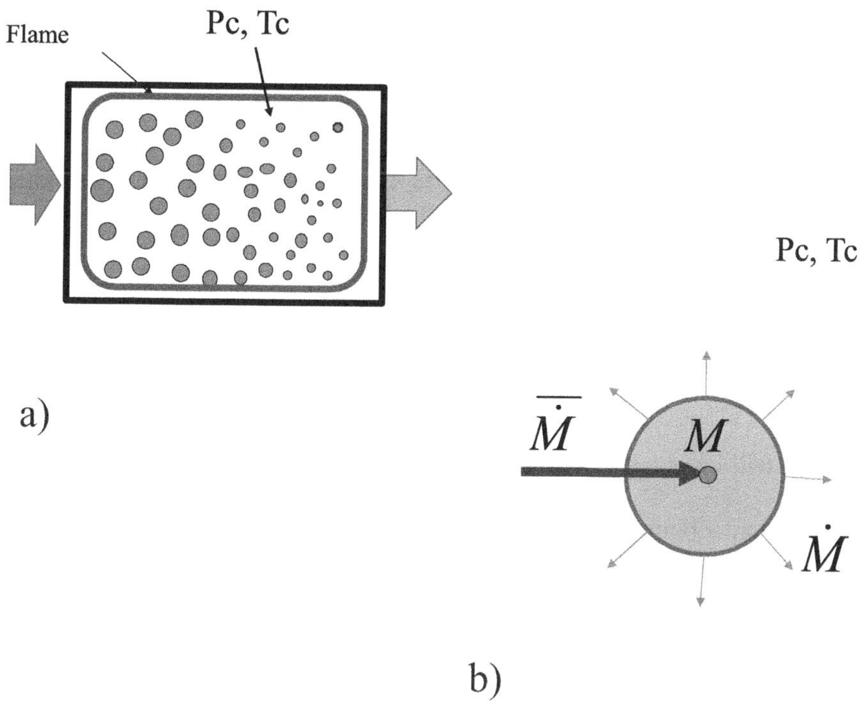

Figure 5.2. *(a) Packet of drops in combustion in a schematized combustion chamber, each drop undergoing evaporation: (left) injection of liquid and gas; (right) output of burned products; within the combustion chamber, we can see a flame that surrounds all the evaporating drops, which become smaller in size towards the right. (b) Individual, continuously fed vaporizing droplet: mass M, evaporation rate $\dot{M}$, the rate of the feed is equal to the mean rate of evaporation $\overline{\dot{M}}$. For a color version of this figure, see www.iste.co.uk/prudhomme/fluid2.zip*

5.3. Equations of the problem

5.3.1. *Equations for the gaseous phase*

According to Spalding's theory, the evaporation rate of a stationary spherical drop can be written as

$$\dot{M} = 4\pi \rho D r_S \ln(1 + B_M)$$

$$B_M = \frac{Y_{FS} - Y_F}{1 - Y_{FS}}, \quad Y_F = \frac{\mathcal{M}_F X_F}{\mathcal{M}_F X_F + \mathcal{M}_A X_A}$$

where the symbols X_j, Y_j and $\mathcal{M}_j$ denote, respectively, the molar fraction, the mass fraction and the molar mass of the species j. B_M is the Spalding mass parameter.

The molar fraction, X_{FS}, of the gaseous F at the surface of the drop is calculated at the evaporation equilibrium, based on the formula for saturated vapor pressure

$$p_C X_{FS} = p_L(T_S)$$

where $p_L(T_S)$ is a known function of the surface temperature T_S. This function will be obtained using the classic formula: $p_L(T_S) = \exp\left(a - \frac{b}{T_S - c}\right)$, with a, b and c being constants.

The heat flow rate communicated to the drop is partly used up in the evaporation of the drop and partly used to heat the interior of the drop. This internal heating Q_L is written in the following form:

$$Q_L = \dot{M}\left(\frac{c_p(T_C - T_S)}{B_T} - l\right) \quad [5.2]$$

with:

$$l = \frac{bRT_S^2}{\mathcal{M}_F(T_s - c)^2}, \text{ and } \rho D \,\text{In}(1 + B_M) = \frac{k}{c_P}\text{In}(1 + B_r) \Rightarrow Le \,\text{In}(1 + B_T)$$

B_T is the Spalding thermal parameter, l is the latent heat of evaporation, Le is the Lewis number, c_p is the heat of the gas per unit mass at a constant pressure, ρ is its density, k is the thermal conduction and D is the coefficient for binary diffusion within the gas.

5.3.2. *Equations for the liquid phase*

In the ***Heidmann and Wieber (1966) model***, the temperature of the drop, T_l, is assumed to be constant and uniform, i.e. the thermal conductivity is infinite. The specific heat of the liquid, c_l, is also constant and uniform and is denoted by c_L. This leads to the following two equations:

$$T_l = T_S$$

$$Mc_L \frac{dT_l}{dt} = Q_L$$

No thermal condition is defined from the point of feeding, which is, in principle, the center of the drop. This is because the drop is isothermal. It is only known that the mass flow rate of the supply feed is equal to the mean evaporation rate of the stabilized (non-perturbed) drop.

Our model of the drop **(Prud'homme *et al.* 2007, 2010; Prud'homme 2010)** continues to assume that the liquid has a constant and uniform specific heat $c_l = c_L$; however, we also assume that the *thermal conductivity is finite and constant* $k_l = k_L$. *We ignore any effect that the convection movements that follow the feeding* may have on the thermal exchanges within the drop.

In the hypothesis where there is spherical symmetry, the only variables are the distance to the center, r, and the time, t. The partial derivative equations and boundary conditions at the surface and at the center of the drop are written as follows, assuming that the thermal gradient is zero at the center of the drop:

$$\frac{\partial(rT_l)}{\partial t} - \kappa_L \frac{\partial^2(rT_l)}{\partial t^2} = 0$$

$$T_l(r_S,t) = T_S, \quad 4\pi r_S^2 k_L \left.\frac{\partial T_l}{\partial r}\right|_{r=r_S} = Q_L \qquad [5.3]$$

$$\left.\frac{\partial T_l}{\partial r}\right|_{r=0} = 0$$

The final condition for $r = 0$ implies a hypothesis where there is an adiabatic center feeding regime in the drop.

5.4. Linearized equations

To study the effects of a small perturbation in the chamber pressure p_C in the liquid and gaseous media, we first linearize the systems of equations.

The following definitions are written for any quantity, f:

– $\bar{f}$: non-perturbed quantity (stabilized reference state);

– $f' = (f - \bar{f})/\bar{f} = \Delta f / \bar{f}$: small relative perturbation.

This second definition is invalid if the reference value for the quantity under consideration is zero, which is the case for the heat, Q_L, which is zero in a stabilized regime where any heat from the gaseous phase is only used for evaporation.

$$\frac{c_p(\bar{T}_C - \bar{T}_S)}{B_T} = l$$

The linearized equations for the gaseous phase are written, when all calculations are made

$$\frac{d\dot{M}'}{dt} + \frac{\dot{M}'}{3\bar{\tau}_v} = \alpha\left(\bar{b}\frac{dT'_S}{dt} - \frac{dp'_C}{dt}\right) \quad [5.4]$$

$$\Delta Q_L = \bar{\dot{M}}\,\bar{l}\,(\bar{a}\,p'_C - \mu T'_S) \quad [5.5]$$

The various coefficients used are explained in Appendix 1 (section 5.8).

For the liquid in the drop, we find, for *infinite thermal conductivity* (the Heidmann case), $\bar{M}\,c_L\bar{T}_S\frac{dT'_S}{dt} = \Delta Q_L$. However, *using our hypotheses*

$$\frac{\partial(rT'_l)}{\partial t} - \kappa_L\frac{\partial^2(rT'_l)}{\partial t^2} = 0$$

$$T'_l(r_S,t) = T'_S,\quad 4\pi r_S^{\,2}k_L\bar{T}_S\left.\frac{\partial T'_l}{\partial r}\right|_{r=r_S} = \Delta Q_L \quad [5.6]$$

$$\left.\frac{\partial T'_l}{\partial r}\right|_{r=0} = 0$$

In this linearized problem, the perturbation in the interior reduced temperature of the drop is equal to the real part of the complex reduced temperature $T'_{lréd}$, and will be denoted by *X*. We thus have:

$X = \Re\left(T'_{l\,réd}\right)$, with: $T'_{l\,réd} = \dfrac{A+B}{B-\theta\,E(u,\theta)} \dfrac{\sinh\left[(1+i)\sqrt{3u/2\theta}\,\xi\right]}{\sinh\left[(1+i)\sqrt{3u/2\theta}\right]} e^{\frac{iu}{3}\tau}$. It is useful to represent X as a function of ξ and τ for different values of u and of θ:

$$\begin{aligned}
&X(u,\theta,\xi,\tau) = \Re\left[T(u,\theta,\xi,\tau)\right] \\
&T(u,\theta,\xi,\tau) = \frac{A+B}{B-\theta E(u,\theta)} \frac{\sinh\left[(1+i)\sqrt{3u/2\theta}\,\xi\right]}{\sinh\left[(1+i)\sqrt{3u/2\theta}\right]} \exp\left(\frac{iu}{3}\tau\right) \qquad [5.12] \\
&E(u,\theta) = 1-(1+i)\sqrt{3u/2\theta}\coth\left[(1+i)\sqrt{3u/2\theta}\right]
\end{aligned}$$

We thus propose tracing the 3D surfaces *(X, ξ, τ)*, over a half-period of the oscillation time, for example, for fixed values of *A*, *B* and *u* and for given values of the parameter θ chosen between 0 and infinity: $A=10, B=100, u=10, \theta=0.1;10;100, 0<\xi<1; \dfrac{\pi}{u}<\tau<\dfrac{7\pi}{u}$.

The representation of the (simplest) case of the drop modeled by a pastille (Anani *et al.* 2011) is given in Appendix 3 (section 5.10) (see also Chapter 4), where we also examine extreme situations with an adiabatic feeding regime and an isothermal regime.

Figure 5.3 is a 3D representation, in the case of a spherical drop with adiabatic feed, of the influence of θ on thermal perturbation as a function of space and time (ξ and τ). It can be seen that at the level of the injection, the spatial slope of the perturbation $\left.\dfrac{\partial T'_l}{\partial r}\right|_{r=0}$ is always zero, as predicted by equation [5.6]. This is the same in the case of the pastille configuration with a zero derivative of space $\left.\dfrac{\partial T'_l}{\partial x}\right|_{x=0}$. At the limit of large values of θ, the thermal perturbation at the injection perfectly follows the thermal perturbation along the surface, with a slight difference in phase. These observations are perhaps even more visible in the 2D representation provided in Appendix 4 (section 5.11).

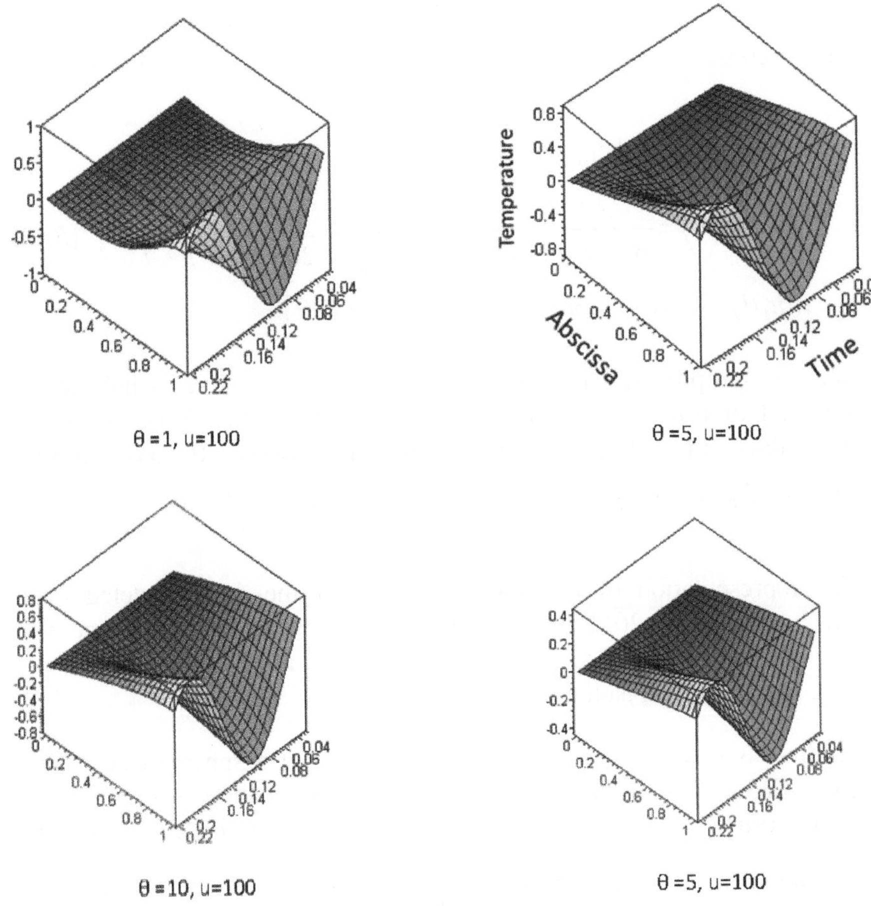

Figure 5.3. *Spherical drop with adiabatic feeding regime: the influence of θ on the thermal perturbation as a function of space and time. For a color version of this figure, see www.iste.co.uk/prudhomme/fluid2.zip*

On the other hand, it can be predicted that in an extreme isothermal feeding regime (Anani *et al.* 2018), the drop will behave like the pastille, i.e. at the level of injection, the thermal perturbation will be zero at all instants, as it is enforced that the liquid that is fed in is at the mean temperature $T_l(0,t) = \overline{T}_L$

5.7. Conclusion

The spherical configuration was studied in a completely analytical manner in the case of an adiabatic feeding regime.

In other cases, we can consider an approximate analytical treatment like that used by Anani *et al.* (2018, ref. vii), which worked successfully in the case of an isothermal feeding regime, taking into account internal convection. However, in the intermediate cases, where the injection is modeled by a simplified law $\left.\dfrac{\partial T_l}{\partial r}\right|_{r=0,t} = \dfrac{h}{r_S}\left(T_l(0,t) - \overline{T}_S\right)$; this is yet to be done and attempts have proven unsuccessful so far, due to the appearance of singularities in the calculation for certain values of the exchange coefficient h. What happens in the process of injection into the center of the drop must indeed be examined in a detailed manner.

5.8. Appendix 1: Coefficients that come into play in linearized equations

$\overline{\tau}_v = \dfrac{\overline{M}}{\dot{\overline{M}}} = \dfrac{\rho_L c_p \overline{r}_S^2}{3k\ln\left(1+\overline{B}_T\right)}$, characteristic time for the passage of the liquid injected into the fed drop: this is also, in a way, the mean lifetime for real drops.

$\widetilde{\tau}_T = \dfrac{\overline{r}_S^2}{9\overline{\kappa}_L}$, characteristic time of thermal conduction.

$\alpha = \dfrac{\overline{B}_M}{\left(1+\overline{B}_M\right)\ln\left(1+\overline{B}_M\right)} \dfrac{\overline{Y}_{AC}\, Y_{FS}}{\overline{Y}_{AS}\left(Y_{FS} - Y_{FC}\right)} \dfrac{\mathcal{M}_F}{\mathcal{M}_F \overline{X}_{FS} + \mathcal{M}_A \overline{X}_{AS}}$, remember that F denotes the liquid in the drop, which appears here in vapor form, A denotes all the other species and is chiefly made up of products of combustion. α is a real number like all the other coefficients that follow.

$\bar{a} = \frac{\bar{T}_C}{\bar{T}_C - \bar{T}_S}\frac{\gamma - 1}{\gamma} + \varphi$, it is assumed that there is isentropic propagation in the gas, which has the coefficient γ.

$$\varphi = \frac{\bar{B}_M}{Le}\frac{\left(1+\bar{B}_M\right)^{\frac{1}{Le}-1}}{\left(1+\bar{B}_T\right)^{\frac{1}{Le}}-1}\frac{\bar{Y}_{AC}\,Y_{FS}}{\bar{Y}_{AS}\left(Y_{FS}-Y_{FC}\right)}\frac{\mathcal{M}_F}{\mathcal{M}_F\bar{X}_{FS}+\mathcal{M}_A\bar{X}_{AS}}$$

$\mu = \frac{\bar{T}_S}{\bar{T}_C - \bar{T}_S} - \frac{2c}{\bar{T}_S - c} + \bar{b}\,\varphi,\ \bar{b} = \frac{b\bar{T}_S}{\left(\bar{T}_S - c\right)^2}$, the coefficients b and c come into play in the expressions for the latent heat of evaporation and the saturated vapor pressure.

5.9. Appendix 2: Solving the thermal equation

The thermal perturbation satisfies equation [5.6]

$$\frac{\partial\left(rT'_l\right)}{\partial t} - \kappa_L\frac{\partial^2\left(rT'_l\right)}{\partial t^2} = 0,$$

with:

$$T'_l\left(r_s,t\right) = T's,\ 4\pi r_s^2 k_L \bar{T}_s \left.\frac{\partial T'_l}{\partial r}\right|_{r=r_s} = \Delta Q_L,\ \left.\frac{\partial T'_l}{\partial r}\right|_{r=0} = 0$$

We must solve a linear partial differential equation with constant coefficients whose elementary solutions are of the form $rT'_l = r\hat{T}_l e^{i\omega t}$.

The function $r\hat{T}_l$ verifies the differential equation $i\omega r\hat{T}_l - \kappa_L\frac{d^2 r\hat{T}_l}{dr^2} = 0$, which admits the general solution $r\hat{T}_l = C\left(e^{s_0 r} - e^{-s_0 r}\right)$ satisfying the condition $\hat{T}_l(0) = 0$. We find the eigenvalue for the characteristic equation $s_0 = (1+i)\sqrt{\omega/2\kappa_L}$. The first boundary condition gives the integration constant $C = r_S \hat{T}_S / \left(e^{s_0 r_S} - e^{-s_0 r_S}\right)$. With the second boundary condition [5.9] at $r = r_S$, $\Delta\hat{Q}_L = -4\pi \bar{r}_S k_L E \hat{T}_S$ and with the approximation $r_S \approx \bar{r}_S$, we obtain $E = 1 - (1+i) s_0 \bar{r}_S \coth s_0 \bar{r}_S$. It can be easily established that $s_0 \bar{r}_S = (1+i)\sqrt{3u/2\theta}$, hence the expression for $E(u,\theta)$ and the expression for Z, given by equation [5.10]. Figure 5.3 corresponds perfectly to the same conditions as in Figure 5.3, however in a 2D representation, which, by definition, is discontinuous over time.

5.10. Appendix 3: The case of the equivalent pastille

Figure 5.4 gives an example of the results obtained for the thermal field of a drop that is non-spherical but shaped like a pastille (Prud'homme and Anani 2018), also called an equivalent drop.

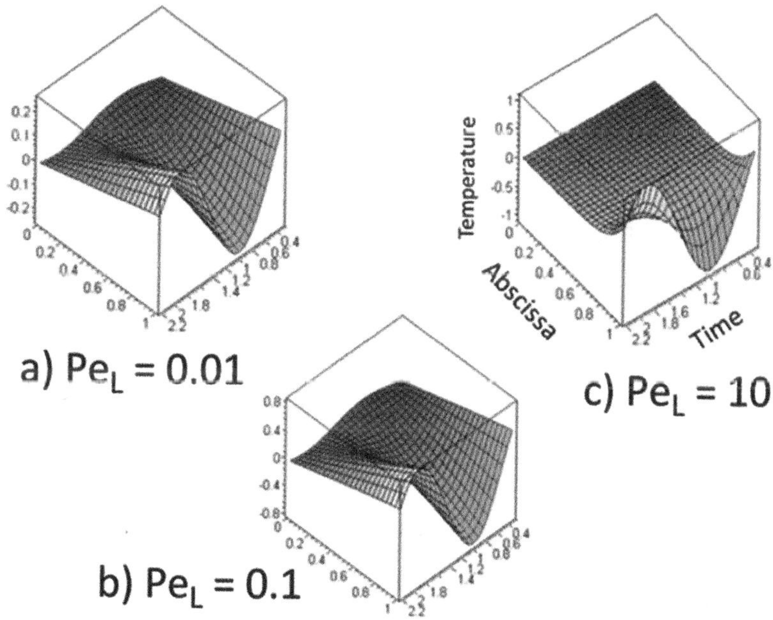

Figure 5.4. *Influence of the Péclet number Pe$_L$ (inverse of θ) with all other parameters having identical values $\varepsilon = 1$, $h = 10$, $u = 10$, $A = 10$, $B = 100$. The axes correspond to the reduced coordinates: temperature X=$T'_{lréd}$, abscissa $\xi = x / x_S$, times $\tau = t / \tau_v$. For a color version of this figure, see www.iste.co.uk/prudhomme/fluid2.zip*

This pastille undergoes a continuous injection of fluid at $x = 0$ with: $\left.\frac{\partial T_l}{\partial x}\right|_{x=0} = \frac{h}{x_S}\left(T_l - \overline{T}_S\right)$, where h is the thermal exchange coefficient at the feed level $x = 0$. The injection is adiabatic if $h = 0$, and is isothermal if h is infinite. This time we take into account the movement of the liquid inside the drop following the feed.

The problem was worked here with an exchange coefficient h equal to 10.

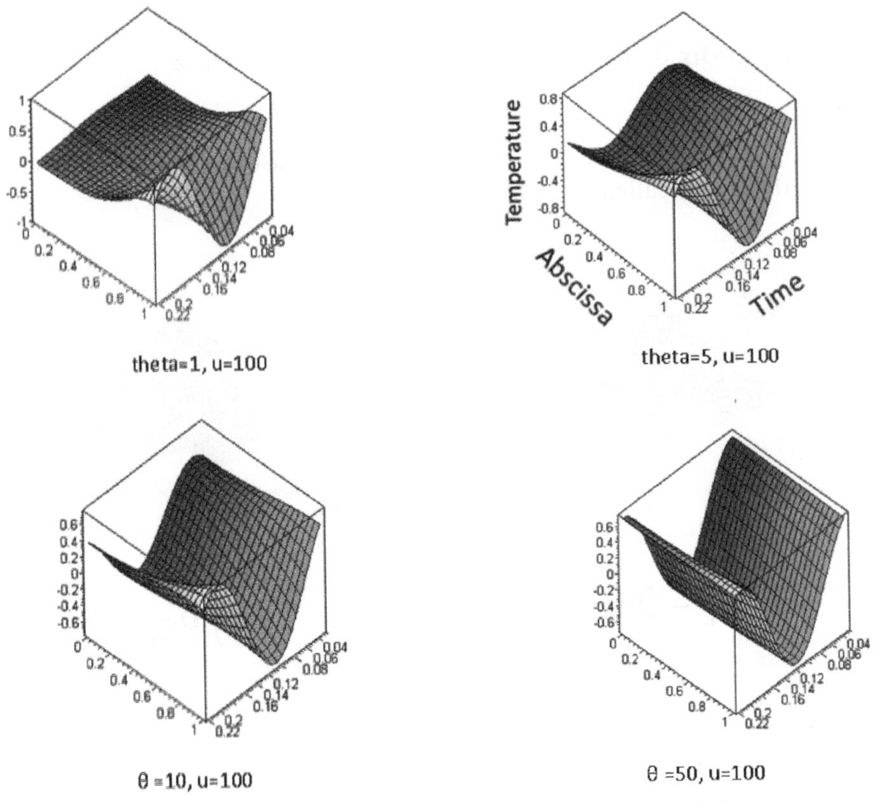

Figure 5.5. *Pastille with adiabatic feeding regime (h = 0). For a color version of this figure, see www.iste.co.uk/prudhomme/fluid2.zip*

In the isothermal case (Figure 5.6), it can be observed that, as predicted, the thermal perturbation is zero for $\xi = 0$ at all instants. It can be predicted that this will be the same case at the center of the spherical drop with an isothermal feeding regime.

Figures 5.5 and 5.6 show the adiabatic and the isothermal cases as a function of θ. It can be seen in the adiabatic case that there is better thermal penetration in the pastille than in the sphere (Figure 5.3). However, the trends are the same, depending on θ.

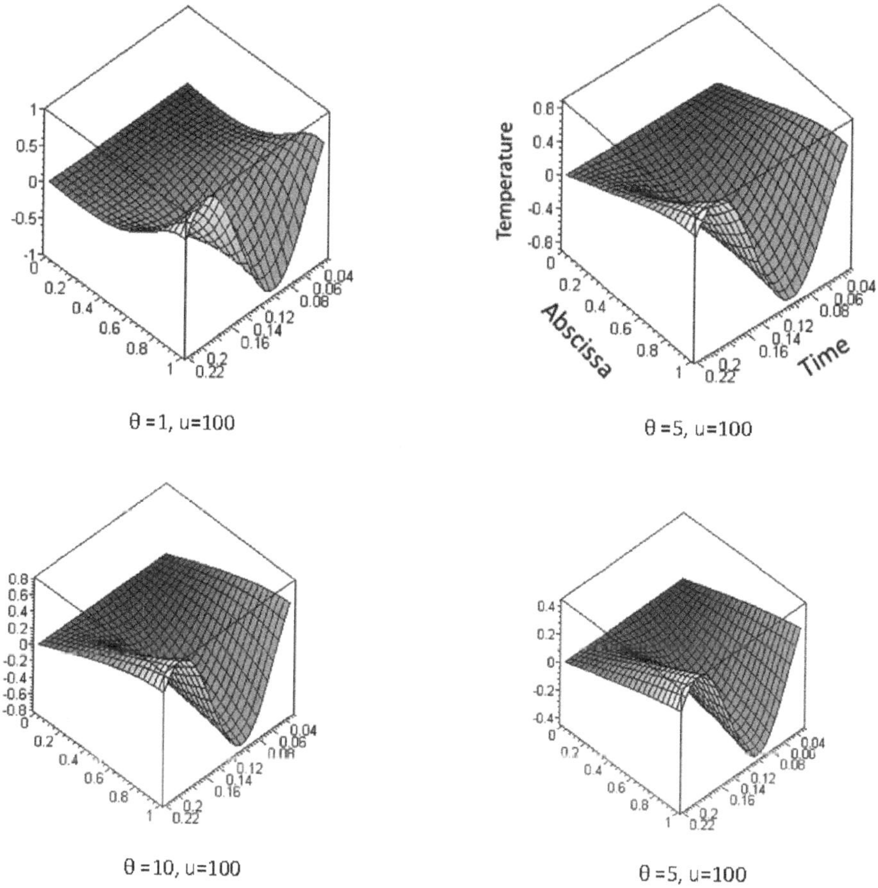

Figure 5.6. *Pastille with isothermal feeding regime (h is infinite). For a color version of this figure, see www.iste.co.uk/prudhomme/fluid2.zip*

5.11. Appendix 4: 2D representation for the spherical drop

Figure 5.7 once again presents – and confirms – the results from Figure 5.3, this time in a 2D representation, for the spherical drop with adiabatic feeding regime.

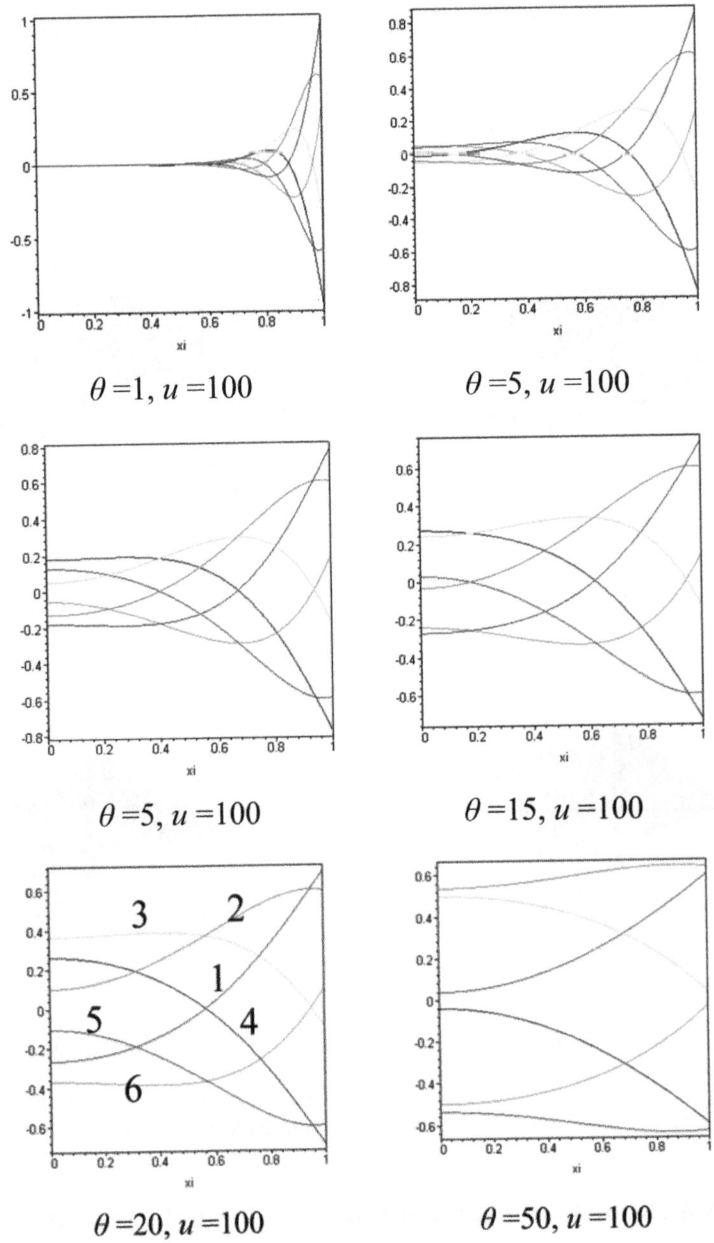

Figure 5.7. *Spherical drop with adiabatic feeding regime: the influence of θ on the thermal perturbation. Each curve corresponds to a reduced time τ for the curves 1 to 6:* $0, \pi/u, 2\pi/u, 3\pi/u, 4\pi/u, 5\pi/u$. *For a color version of this figure, see www.iste.co.uk/prudhomme/fluid2.zip*

5.12. References

Anani, K., Prud'homme, R., D'Almeida, A., Assiamoua, K. (2011). Effect of thermal convection on frequency response of a perturbed vaporizing pastille-shaped droplet. *Mécanique & industries*, 12(4), 301–314.

Anani, K., Prud'homme, R., Hounkonnou, M.N. (2018). Dynamic response of a vaporizing spray to pressure oscillations: Approximate analytical solutions. *Combustion and Flame*, 193, 295–305 [Online]. Available at: https://www.sciencedirect.com/science/article/abs/pii/S0010218018301251.

Heidmann, M.F. (1972). Frequency response of a vaporization process to distorted acoustic disturbances. NASA Technical Note, D-6806.

Heidmann, M.F. and Wieber, P.R. (1966). Analysis of frequency response characteristics of propellant vaporization. NASA Technical Note, D-3749.

Prud'homme, R. (2010). *Flows of Reactive Fluids*. Springer, New York, Dordrecht, Heidelberg, London.

Prud'homme, R. and Anani K. (2018). Évaporation d'une pastille équivalente. *ISTE OpenScience*, 2(1) [Online]. Available at: https://www.openscience.fr/Vaporization-of-an-equivalent-pastille.

Prud'homme, R., Mauriot, Y., Habıballah, M. (2007). Modelling of internal exchanges of a vaporizing droplet submitted to pressure oscillations. *6th REST Scientific Workshop*, Munich, Germany, 29–31 March [Online]. Available at: https://www.researchgate.net/publication/273383677_Modelling_of_internal_exchanges_of_a_vaporizing_droplet_submitted_to_pressure_oscillations.

Prud'homme, R., Habiballah, M., Matuszewski, L., Mauriot, Y., Nicole, A. (2010). Theoretical analysis of dynamic response of a vaporizing droplet to acoustic oscillations. *Journal of Propulsion and Power*, 26(1), 74–83.

6

Study of the Three-Dimensional and Non-Stationary Flow in a Rotor of the Savonius Wind Turbine

The purpose of this study is to develop a computational code for solving the equations governing the flow of fluids in a vertical-axis wind turbine. In order to check the effectiveness of our code, we carried out an application on the Savonius rotor. The equations that define the flow of fluids in a rotating wheel are continuity equations, Navier–Stokes and energy equations. These equations are solved numerically by the explicit Lax–Wendroff finite difference method, followed by the addition of a time-corrected artificial viscosity. The schema is second-order accurate with respect to space-time, and the stability of the numerical computation is ensured by the CFL condition, which imposes a constraint on the time step. The results are presented through pressure and Mach curves.

6.1. Introduction

Wind energy has become the most effective alternative energy source, especially in the industrial and energy sectors. Several research studies have been carried out to evaluate the performance of a horizontal-axis wind turbine [NOU 15, SOS 18]. Although vertical-axis wind turbines have a low output [ZHA 17, GHA 14, WID 12], there are studies being carried out to determine how their functioning can be improved [SON 16, ELK 15]. Among the various studies that have already examined this question, some were experimental in nature, using wind tunnels and measuring instruments. These varied the number of blades [HAD 13] or the shape [SOS 18]; however, they

Chapter written by Francis RAVELOSON, Delphin TOMBORAVO and Roger VONY.

all required a very powerful laboratory. Other studies were numerical and used CFD software [DOB 12, MAR 17, WAH 13], which was predefined, like Ansys-Fluent [KAN 13, DEB 13, DWI 16], "SolidWorks Flow Simulation" [WID 12], or again MCC+ [BUT 13], in order to see how to make a wind turbine profitable. However, these are all commercial software and are very expensive, which prevents some third-world countries from using them. Further, in most cases the Fluent software is used to numerically solve URANS equations [NOU 15]. Thus, a purely 3D approach using free-software to determine the fine flow structure through the blades of a vertical-axis wind turbine is still very uncommon.

Given these problems, the current study carries out a numerical simulation of a 3D, non-stationary flow in a Savonius, vertical-axis wind turbine. The simulation is carried out using the free software Fortran to study the aerodynamic structure of this rotor. This numerical simulation is carried out using the explicit discretization method, for which the Lax scheme with two time steps has been used, in order to enhance the accuracy of the results. Further, the CFD simulation is used to calculate the aerodynamic effort, as well as the mechanical torque recorded at the shaft runout of this sensor.

6.2. Mathematical modeling of the problem

6.2.1. *Presentation of a physical model*

The geometry of the calculation domain and that of a Savonius wind turbine are presented in Figure 6.1. The geometric data that we used in this study is as follows: overlap e = 15 cm, radius R = 50 cm and height H = 120 cm.

For the movement of a fluid to be mathematically described, it is considered as a continuous medium. It therefore obeys the fundamental principles of mechanics and thermodynamics. Three frames are used in this study:

– the absolute frame formed by three unit vectors $\vec{\imath}$, $\vec{\jmath}$ and $\vec{k}$;

– the relative orthonormal frame denoted by (E), rotating around an axis defined by three unit vectors attached to a point M, located in the flow (M, $\vec{e}_m$, $\vec{e}_y$, $\vec{e}_n$);

– another mobile frame called (H), defined by three unit vectors $\vec{h}_m$, $\vec{h}_n$ and $\vec{h}_y$, attached to the points of the meshing (pseudo-current nappe).

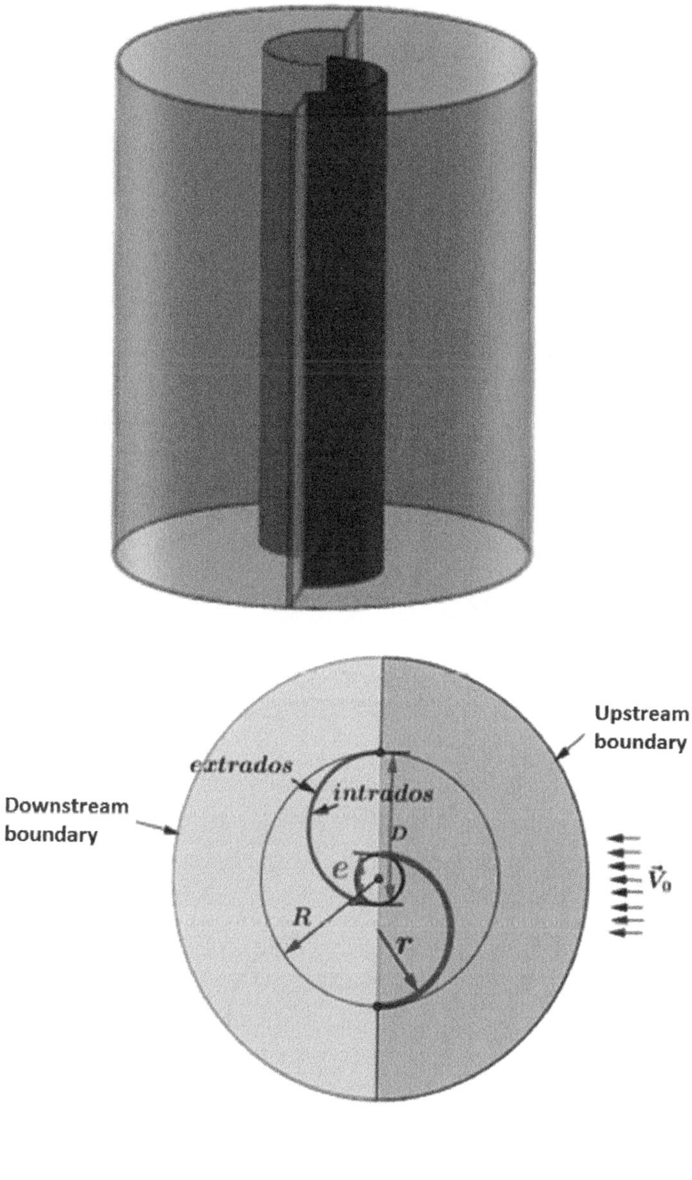

Figure 6.1. *Description of a Savonius rotor and the calculation field. For a color version of this figure, see www.iste.co.uk/prudhomme/fluid2.zip*

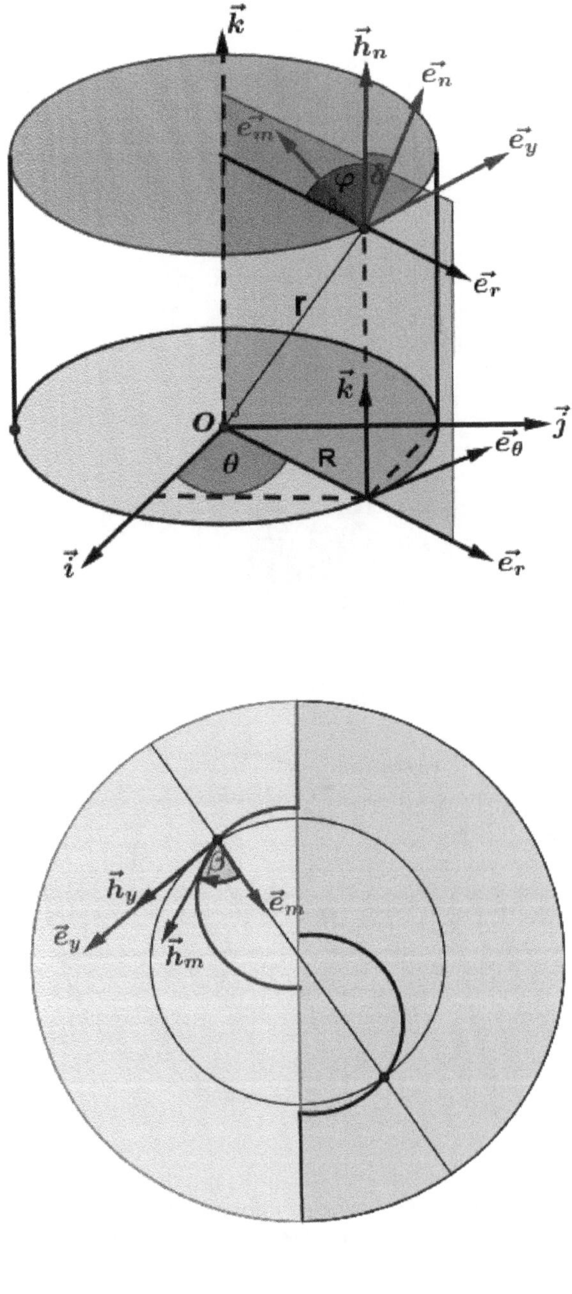

Figure 6.2. *Presentation of the three frames used. For a color version of this figure, see www.iste.co.uk/prudhomme/fluid2.zip*

6.2.2. *Simplifying hypotheses*

In this study, we have used the following simplifying hypotheses in order to better solve the problems without too much variation in the physical phenomena that exist:

– the fluid is a perfect gas;

– the fluid is weightless;

– thermal conductivity is non-negligible.

Taking into account these simplifying hypotheses, the movement equations in the mobile frame (H) are written:

– continuity equation:

$$\frac{\partial \rho'}{\partial t} = W_m \frac{\partial \rho'}{\partial m} + \varepsilon \frac{\partial \rho'}{\partial y} + W_n \frac{\partial \rho'}{\partial n} + \frac{\partial W_m}{\partial m} + \frac{\partial W_y}{\partial m} - tan\beta \frac{\partial W_m}{\partial y} - tan\delta \frac{\partial W_n}{\partial y} + \frac{\partial W_n}{\partial n} - W_m \left(\frac{sin\varphi}{r} + \frac{\partial \varphi}{\partial n}\right) + W_n \left(\frac{cos\varphi}{r} + \frac{\partial \varphi}{\partial m}\right) \quad [6.1]$$

– equation of conservation of momentum (Euler's equation):

$$\frac{\partial W_m}{\partial t} = W_m \frac{\partial W_m}{\partial m} + \varepsilon \frac{\partial W_m}{\partial y} + W_n \frac{\partial W_m}{\partial n} + R_g \left[\frac{\partial T}{\partial m} - tan\beta \frac{\partial T}{\partial y}\right] + \frac{a^2}{\gamma}\left[\frac{\partial \rho'}{\partial m} - tan\delta \frac{\partial \rho'}{\partial y}\right] + \frac{{W_y}^2}{r} sin\varphi + W_n \left(W_n \frac{\partial \varphi}{\partial n} + W_m \frac{\partial \varphi}{\partial m}\right) \quad [6.2]$$

$$\frac{\partial W_y}{\partial t} = W_m \frac{\partial W_y}{\partial m} + \varepsilon \frac{\partial W_y}{\partial y} + W_n \frac{\partial W_y}{\partial n} + R_g \frac{\partial T}{\partial m} + \frac{a^2}{\gamma} \frac{\partial \rho'}{\partial m} + (W_y + 2U)\left(W_n \frac{cos\varphi}{r} + W_m \frac{sin\varphi}{r}\right) \quad [6.3]$$

$$\frac{\partial W_n}{\partial t} = W_m \frac{\partial W_n}{\partial m} + \varepsilon \frac{\partial W_n}{\partial y} + W_n \frac{\partial W_n}{\partial n} + R_g \left[\frac{\partial T}{\partial n} - tan\beta \frac{\partial T}{\partial y}\right] + \frac{a^2}{\gamma}\left[\frac{\partial \rho'}{\partial n} - tan\delta \frac{\partial \rho'}{\partial y}\right] + \frac{{W_y}^2}{r} cos\varphi - W_m \left(W_n \frac{\partial \varphi}{\partial n} + W_m \frac{\partial \varphi}{\partial m}\right) \quad [6.4]$$

– energy equation:

$$\frac{\partial T}{\partial t} = \frac{a^2}{C_p}\left[\frac{\partial W_m}{\partial m} - tan\beta \frac{\partial W_m}{\partial y} + \frac{\partial W_y}{\partial y} + \frac{\partial W_n}{\partial n} - tan\delta \frac{\partial W_n}{\partial y} - W_m \left(\frac{sin\varphi}{r} + \frac{\partial \varphi}{\partial n}\right) + W_n \left(\frac{cos\varphi}{r} + \frac{\partial \varphi}{\partial m}\right)\right] + W_m \frac{\partial T}{\partial m} + \varepsilon \frac{\partial T}{\partial y} + W_n \frac{\partial T}{\partial n} \quad [6.5]$$

6.3. Numerical resolution

The majority of the methods discussed in literature for solving Euler equations are explicit methods, the best known of which are the Lax–Wendroff and MacCormack methods. These are very convenient in solving non-stationary problems as the schemes are second-order accurate with respect to time and space. The resolution method presented in this article is the Lax–Wendroff method, which was largely used for calculating flows in turbomachines.

6.3.1. *Presentation of meshes*

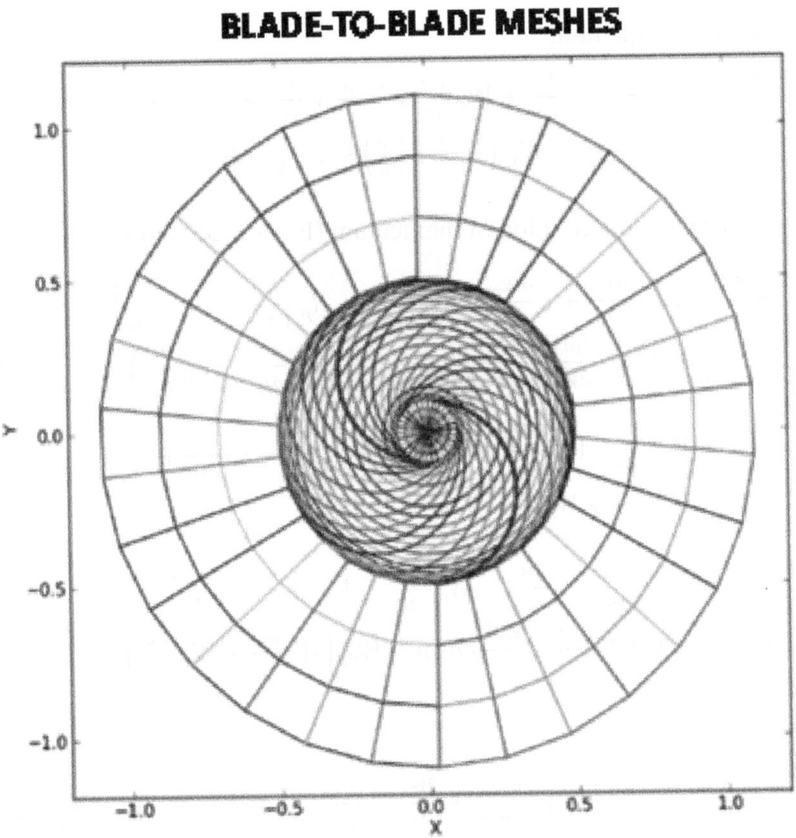

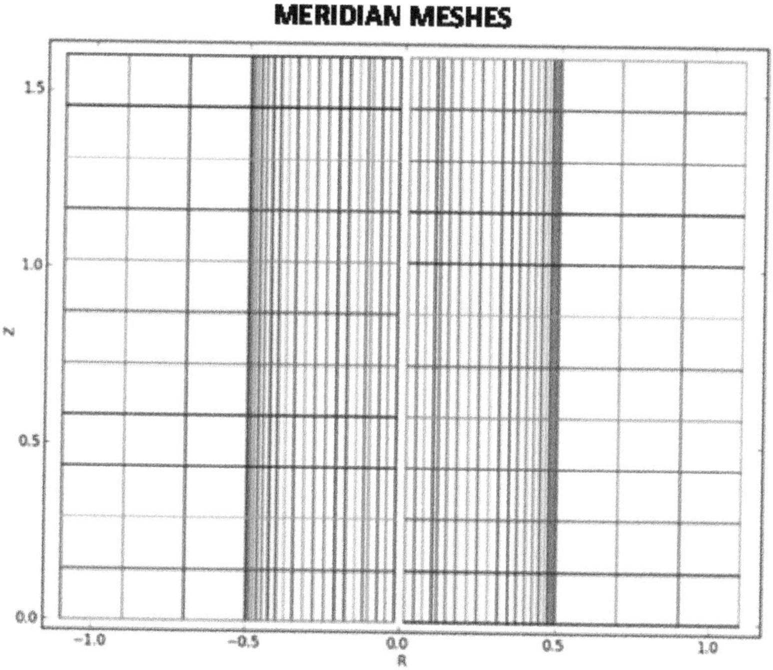

Figure 6.3. *Meshes in the blade-to-blade plane and meridian plane. For a color version of this figure, see www.iste.co.uk/prudhomme/fluid2.zip*

We often use orthogonal meshes associated with the "finite difference" method over the domain of numerical simulation of fluids governed by the Navier–Stokes equation [PRU 18]. This is the case with the present study, which generates a 3D orthogonal mesh. Here, the floor of each of the blades or arms corresponds to a current sheet. The meshing steps are more relaxed outside of the inter-blade channel and are finer in the neighborhood of the leading edge and the trailing edge. In order to avoid information overload in Figure 6.4, we have only presented a 3D mesh at the level of the foot, halfway up, at the head, as well as a mesh of the model in Figure 6.1. The mesh in the meridian plane is made up of generatrices and lines that are perpendicular to them. In the blade-to-blade plane, the mesh is made up of arcs of concentric circles whose center coincides with the machine's axis of rotation.

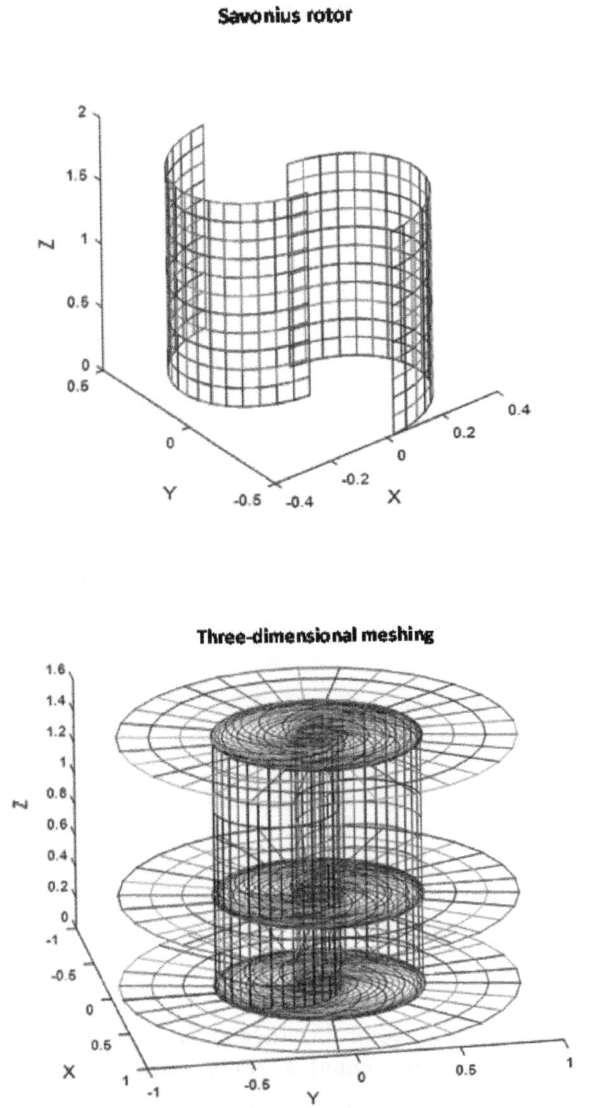

Figure 6.4. *Meshing on the blades and at the foot, halfway-up and at the head of the blades. For a color version of this figure, see www.iste.co.uk/prudhomme/fluid2.zip*

6.3.2. *Spatial discretization*

The problem consists of moving from a continuous structure to an equivalent discretized structure [KAN 18]. Two cases must be considered for spatial discretization: the case of points located in the interior of the domain, and that of points situated on the borders of the calculation domain. The centered schema is applied to the points located in the interior of the calculation domain and is presented as follows:

$$\left[\frac{\partial \vec{Z}}{\partial S}\right]_{(l)} = \frac{\nabla S}{\Delta S(\nabla S+\Delta S)}\vec{Z}^{\tau}{}_{(l+1)} + \frac{\Delta S-\nabla S}{\Delta S\,\nabla S}\vec{Z}^{\tau}{}_{(l)} - \frac{\Delta S}{\Delta S(\nabla S+\Delta S)}\vec{Z}^{\tau}{}_{(l-1)} \qquad [6.6]$$

We have used a schema that is decentered towards the interior for the case where points are placed along the boundaries. Along the upstream boundaries, to the left and at the feet of the blades, the spatial discretization is written as

$$\left[\frac{\partial \vec{Z}}{\partial S}\right]_{(l)} = \frac{\nabla S}{\Delta S(\nabla S+\Delta S)}\vec{Z}^{\tau}{}_{(l+2)} + \frac{\Delta S-\nabla S}{\Delta S\,\nabla S}\vec{Z}^{\tau}{}_{(l+1)} - \frac{\Delta S}{\Delta S(\nabla S+\Delta S)}\vec{Z}^{\tau}{}_{(l)} \qquad [6.7]$$

While on the downstream boundaries, to the right and at the head of the blades, it becomes:

$$\left[\frac{\partial \vec{Z}}{\partial S}\right]_{(l)} = \frac{\nabla S}{\Delta S(\nabla S+\Delta S)}\vec{Z}^{\tau}{}_{(l)} + \frac{\Delta S-\nabla S}{\Delta S\,\nabla S}\vec{Z}^{\tau}{}_{(l-1)} - \frac{\Delta S}{\Delta S(\nabla S+\Delta S)}\vec{Z}^{\tau}{}_{(l-2)} \qquad [6.8]$$

In equations [6.6]–[6.8], $\vec{Z}$ replaces the variables to be discretized, S replaces the spatial direction, l replaces the index corresponding to the direction being considered and ∇S and ΔS are, respectively, the mesh cell to the left and right of the point being studied, except along the borders.

6.3.3. *Temporal discretization*

A variant of the Lax two time step scheme was applied for temporal discretization. The first step of this scheme corresponds to the Lax–Friedrichs (predictor) and the Lax–Wendroff for the second step (corrector). A numerically corrected viscosity scheme is also applied to the third calculation step including:

– Lax–Friedrichs (predictor) scheme

$$\vec{Z}^{\tau+1}_{(i,j,k)} = \frac{1}{6}\left(\vec{Z}^{\tau}_{(i+1,j,k)} + \vec{Z}^{\tau}_{(i-1,j,k)} + \vec{Z}^{\tau}_{(i,j+1,k)} + \vec{Z}^{\tau}_{(i,j-1,k)} + \vec{Z}^{\tau}_{(i,j,k+1)} + \vec{Z}^{\tau}_{(i,j,k-1)}\right) + \Delta t * \left(B^{\tau}_{m(i,j,k)}\left[\frac{\partial \vec{Z}}{\partial m}\right]^{\tau}_{(i,j,k)} + B^{\tau}_{y(i,j,k)}\left[\frac{\partial \vec{Z}}{\partial y}\right]^{\tau}_{(i,j,k)} + B^{\tau}_{n(i,j,k)}\left[\frac{\partial \vec{Z}}{\partial n}\right]^{\tau}_{(i,j,k)} + C^{\tau}_{(i,j,k)}\right) \quad [6.9]$$

where τ is the temporal discretization index, and B_m, B_y, B_n and C are the variables to be discretized depending on the direction considered, represented in the form of a matrix.

– Lax–Wendroff scheme (corrector)

$$\vec{Z}^{\tau+2}_{(i,j,k)} = \vec{Z}^{\tau}_{(i,j,k)} + \Delta t * \left(B^{\tau+1}_{m(i,j,k)}\left[\frac{\partial \vec{Z}}{\partial m}\right]^{\tau+1}_{(i,j,k)} + B^{\tau+1}_{y(i,j,k)}\left[\frac{\partial \vec{Z}}{\partial y}\right]^{\tau+1}_{(i,j,k)} + B^{\tau+1}_{n(i,j,k)}\left[\frac{\partial \vec{Z}}{\partial n}\right]^{\tau+1}_{(i,j,k)} + C^{\tau+1}_{(i,j,k)}\right) \quad [6.10]$$

This discretization schema is dissipative: it carries depreciation terms inside it, but these are not enough to cause the calculation to converge. Thus, it is necessary to add a corrected numerical viscosity, which reinforces the dissipative effects and accelerates the convergence of the calculations. However, this artificial viscosity will disappear at the end of the calculation and the error it generates will not affect the precession of the results.

– Scheme with corrected numerical viscosity

$$\vec{Z}^{\tau+3}_{(i,j,k)} = \vec{Z}^{\tau+2}_{(i,j,k)} - VN * \left(\vec{Z}^{\tau+2}_{(i,j,k)} - e^{*}{}_{(i,j,k)}\,\vec{Z}^{*}_{(i,j,k)}\right) \quad [6.11]$$

with $e^{*}{}_{(i,j,k)}$ being the numerical viscosity coefficient and $\vec{Z}^{*}_{(i,j,k)}$ the variable obtained at the intermediate instant.

6.3.4. *Stability condition for the scheme*

The condition for the stability of the scheme is dictated by the CFL (Courant–Friedrichs–Lewy) condition:

$$\Delta t < \frac{h_{(i,j,k)}}{\sqrt{3}(W+a)_{(i,j,k)}} \quad [6.12]$$

with $h_{(i,j,k)}$ being the smallest meshing cell in a calculation domain, W the resultant of the relative velocity and a the modulus of the speed of sound.

6.3.5. *Initial conditions*

At the initial instant and across the section that is perpendicular to the principal direction of flow ($\vec{e}_m$):

– the flow rate remains constant: $Q = \rho \, W_m A = Q_1$;

– the rothalpy remains constant: $I = \frac{W^2}{2} C_p \, T + \frac{\omega^2 \, r^2}{2} = C^{te}$;

– the entropy is assumed to be constant along the lines of the mesh (no irreversibility): $\frac{\rho}{\rho_1} = \left(\frac{T}{T_1}\right)^{\frac{1}{\gamma-1}}$.

6.3.6. *Boundary conditions*

These are the boundary conditions that we adopted during the numerical solution.

Upstream boundary:

– the absolute velocity and the angle indicating its direction are fixed;

– the temperature is equal to the static temperature.

Downstream boundary: atmospheric pressure.

Lateral boundary: condition of equality of all variables.

Solid walls (blades): sliding condition.

At the foot and head of the blades: condition of non-penetration.

6.4. Validation of the results

Before we present our results of the numerical simulation, it is advisable to validate the calculation code that we designed, either through experimental study or through a numerical analysis that is already validated. In this case, our code will be validated through the study carried out by Butaud *et al.* [BUT 13]. The geometric and simulation characteristics of this study are specified below. This was on a classic Savonius wind turbine, 1 m in diameter and an optimal sweep of 0.15; its semi-cylindrical aluminum blades are 3 mm thick. They carried out the 2D numerical simulation using the Star-CCM+ software, with the chosen time step of 10^{-3} s and the number of rotor rotations required for the proper dynamic stabilization of the system being fixed to 12 rounds. The fluid is viscous, with a high Reynolds number (of the order of 5.10^6), which naturally results in a URANS (k - ω) model. The team carried out an unstable numerical study with the wind velocity at the inlet to the domain being set to 7 m/s. The boundary conditions at the edges were symmetrical conditions, atmospheric pressure at the outlet and wall condition on the blades. The dynamic pressure distribution for the angular position $\theta = 0°$ will be compared to our 3D studies.

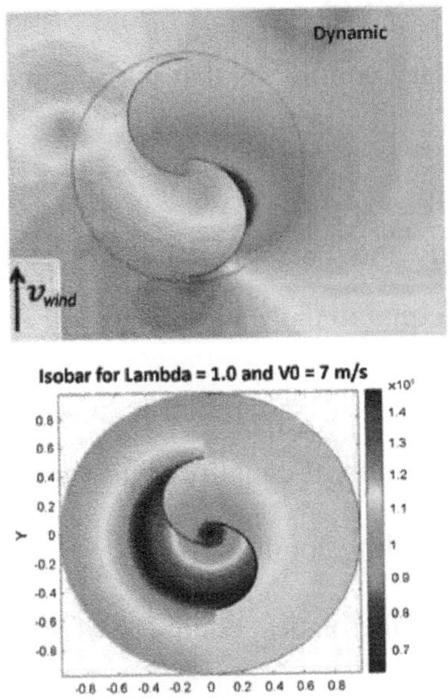

Figure 6.5. *Dynamic pressure depression in the study by Butaud (top); and that of the current study presented in this text (bottom). For a color version of this figure, see www.iste.co.uk/prudhomme/fluid2.zip*

6.5. Results and discussion

6.5.1. *Influence of the advance parameter*

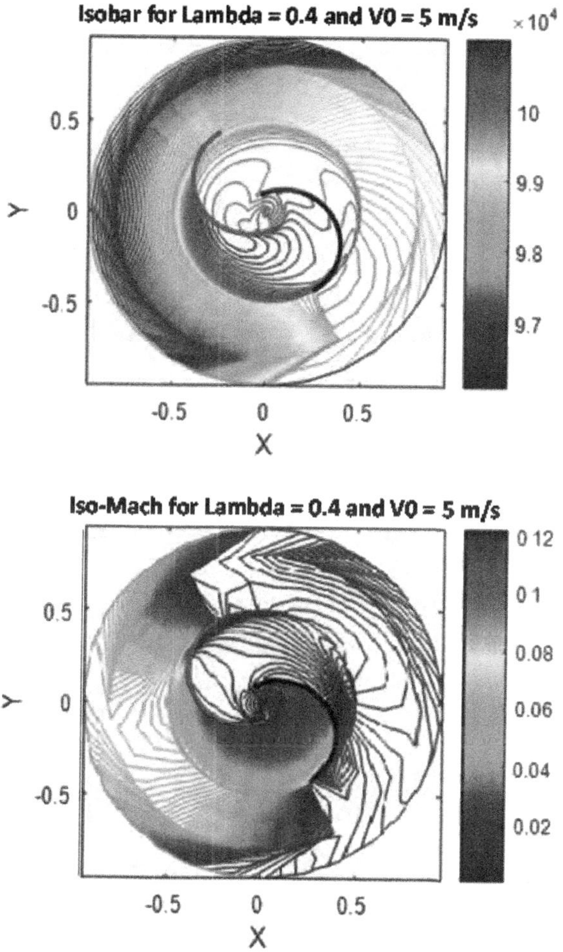

Figure 6.6. *Isobar and iso-Mach curves for* λ *= 0.4. For a color version of this figure, see www.iste.co.uk/prudhomme/fluid2.zip*

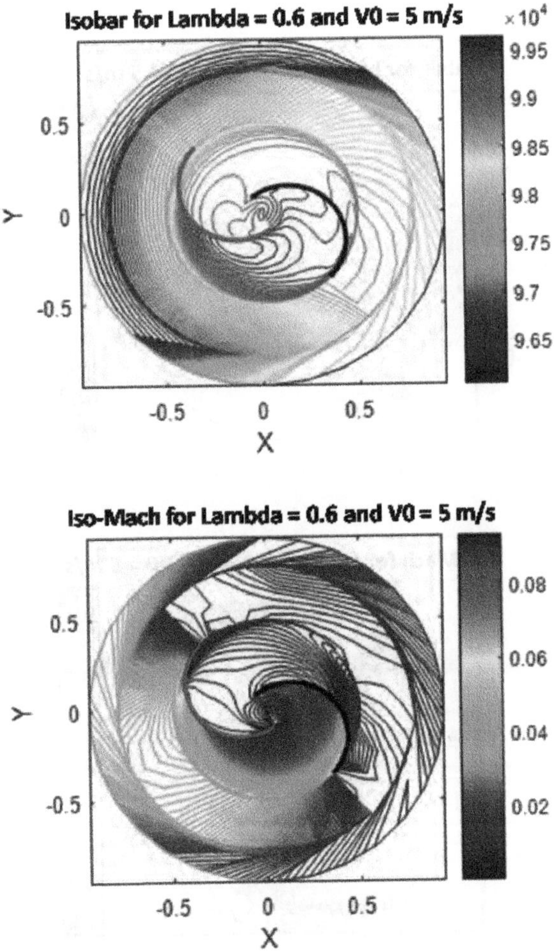

Figure 6.7. *Isobar and iso-Mach curves for λ = 0.6. For a color version of this figure, see www.iste.co.uk/prudhomme/fluid2.zip*

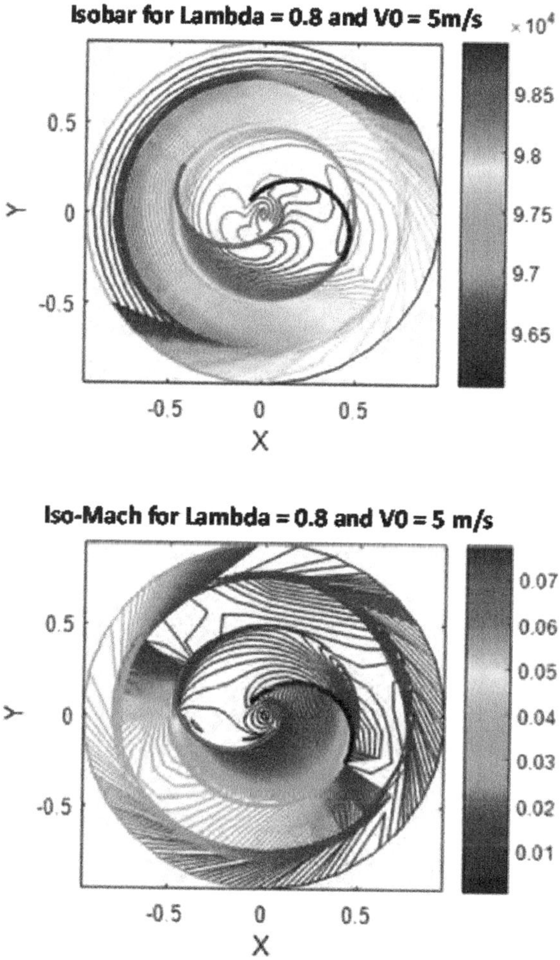

Figure 6.8. *Isobar and iso-Mach curves for $\lambda = 0.8$. For a color version of this figure, see www.iste.co.uk/prudhomme/fluid2.zip*

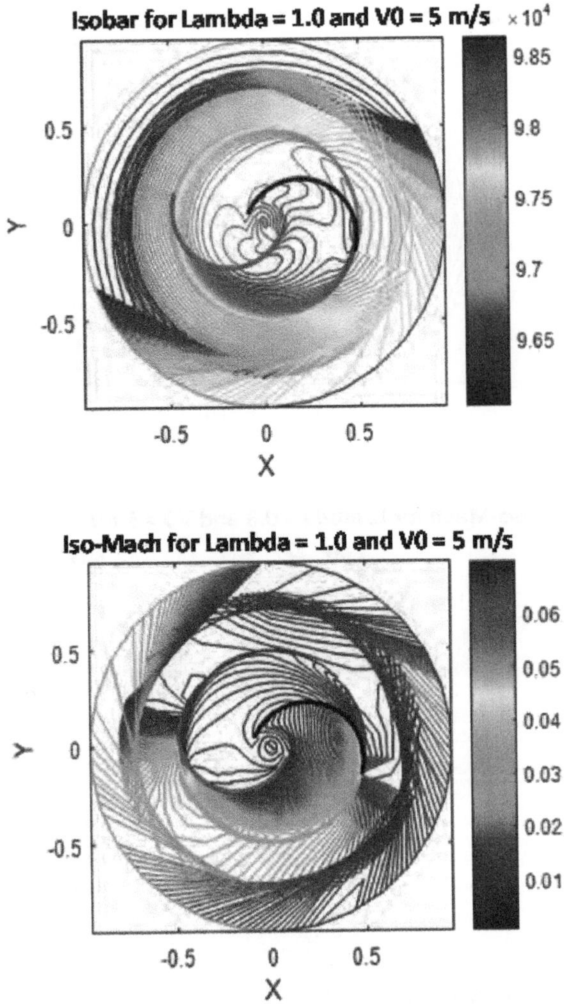

Figure 6.9. *Isobar and iso-Mach curves for $\lambda = 1.0$. For a color version of this figure, see www.iste.co.uk/prudhomme/fluid2.zip*

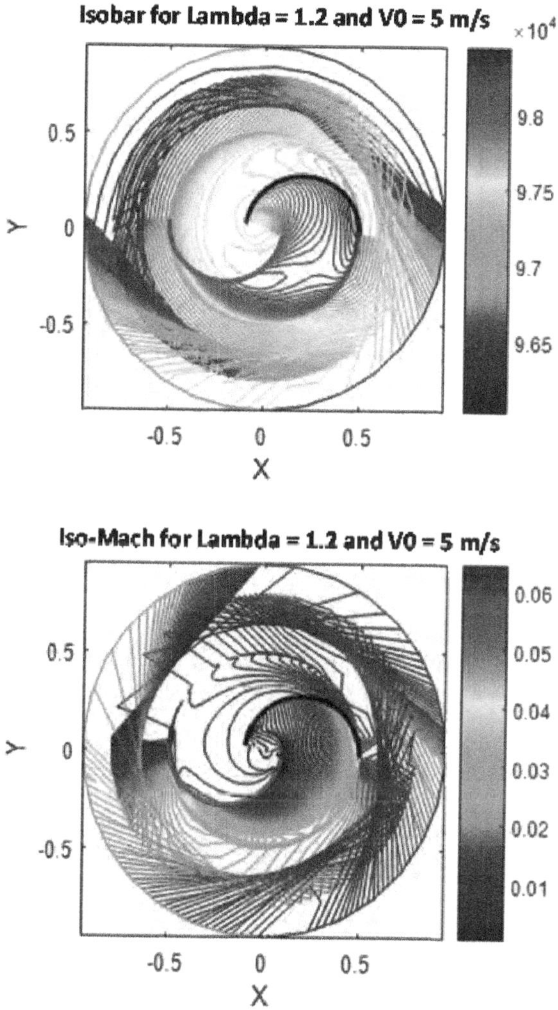

Figure 6.10. *Isobar and iso-Mach curves for λ = 1.2. For a color version of this figure, see www.iste.co.uk/prudhomme/fluid2.zip*

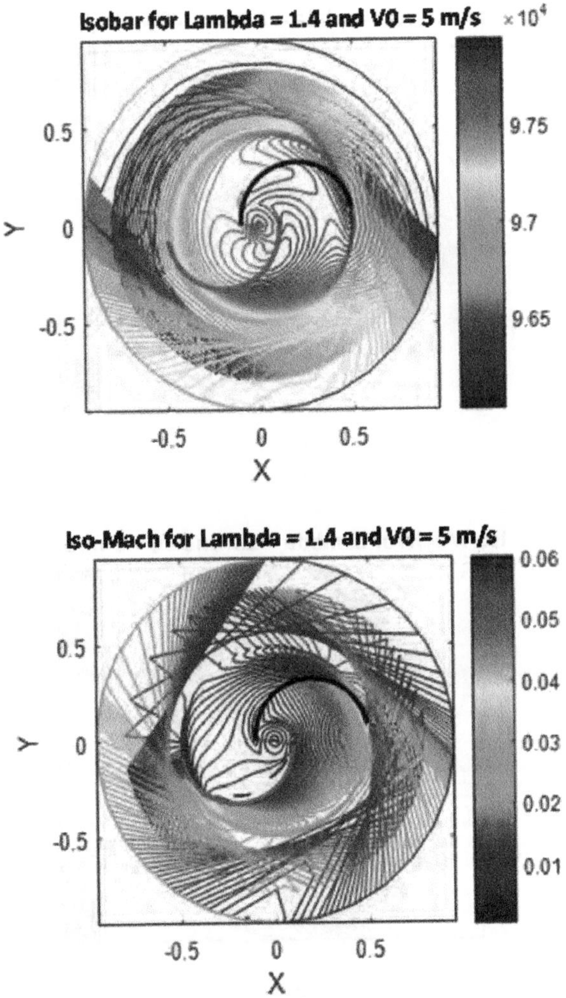

Figure 6.11. *Isobar and iso-Mach curves for λ = 1.4. For a color version of this figure, see www.iste.co.uk/prudhomme/fluid2.zip*

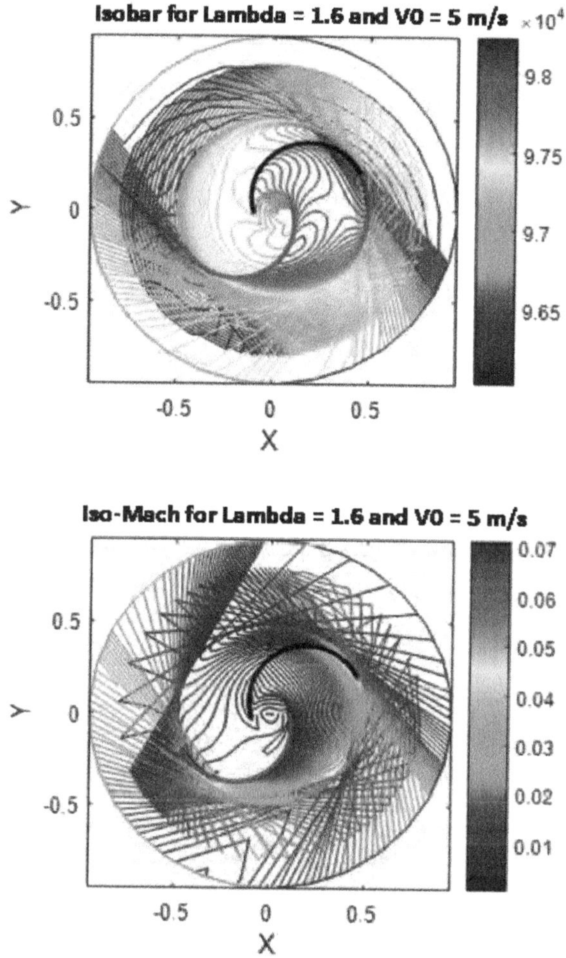

Figure 6.12. *Isobar and iso-Mach curves for $\lambda = 1.6$. For a color version of this figure, see www.iste.co.uk/prudhomme/fluid2.zip*

6.5.2. *Influence of the angular position of the blades*

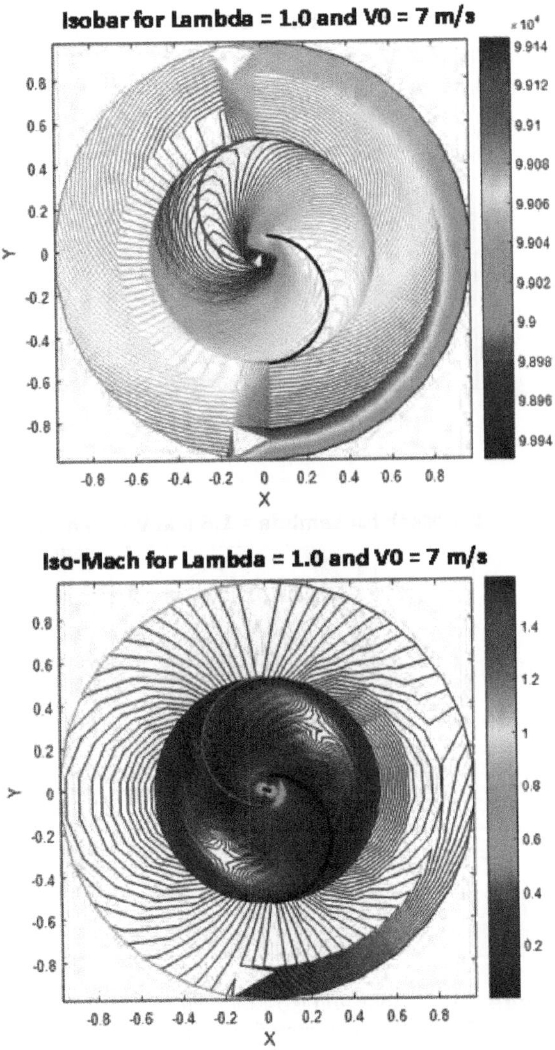

Figure 6.13. *Isobar and iso-Mach curves at t = 0 s and α = 90°. For a color version of this figure, see www.iste.co.uk/prudhomme/fluid2.zip*

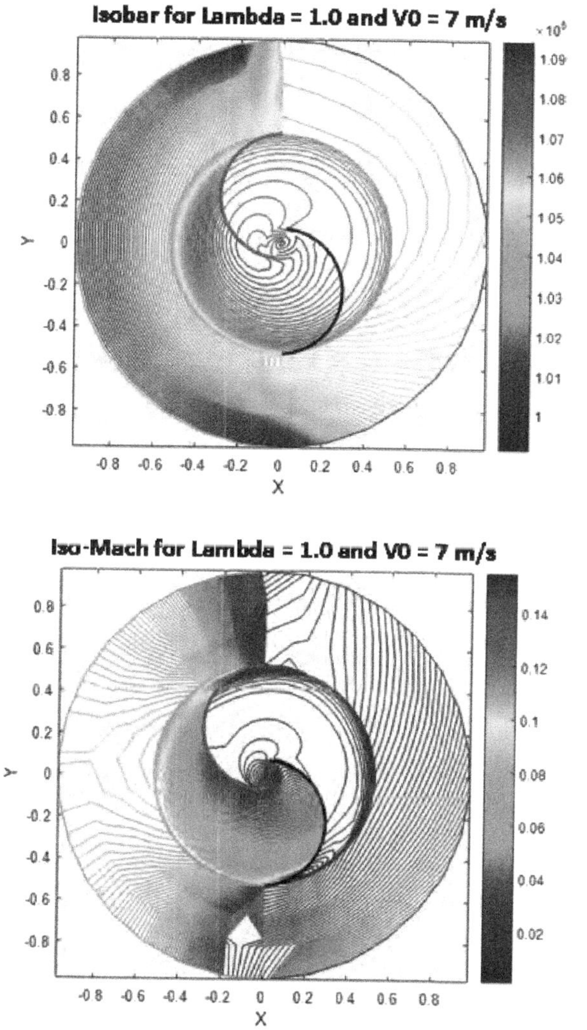

Figure 6.14. *Isobar curves at the instant t = 0.927 s, α = 90° after one rotation. For a color version of this figure, see www.iste.co.uk/prudhomme/fluid2.zip*

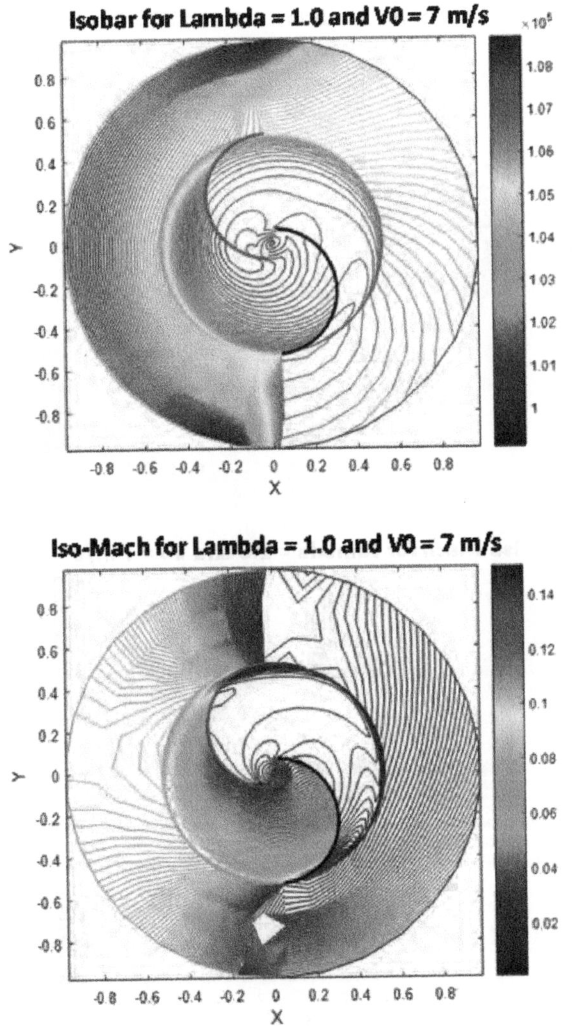

Figure 6.15. *Isobar curves at the instant t = 5.677 s, α = 90° after six rotations. For a color version of this figure, see www.iste.co.uk/prudhomme/fluid2.zip*

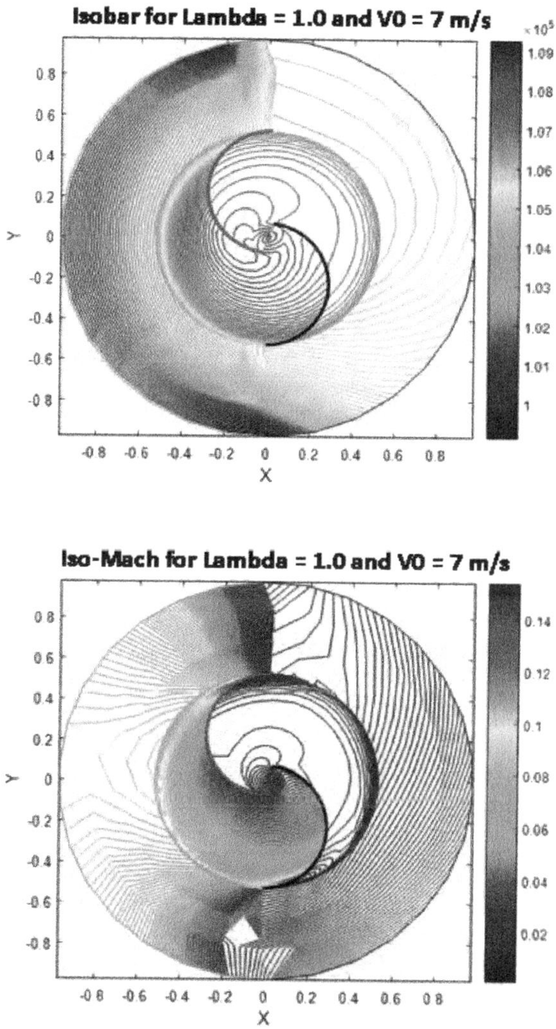

Figure 6.16. *Isobar curves at the instant t = 11.35 s, α = 90° after 12 rotations. For a color version of this figure, see www.iste.co.uk/prudhomme/fluid2.zip*

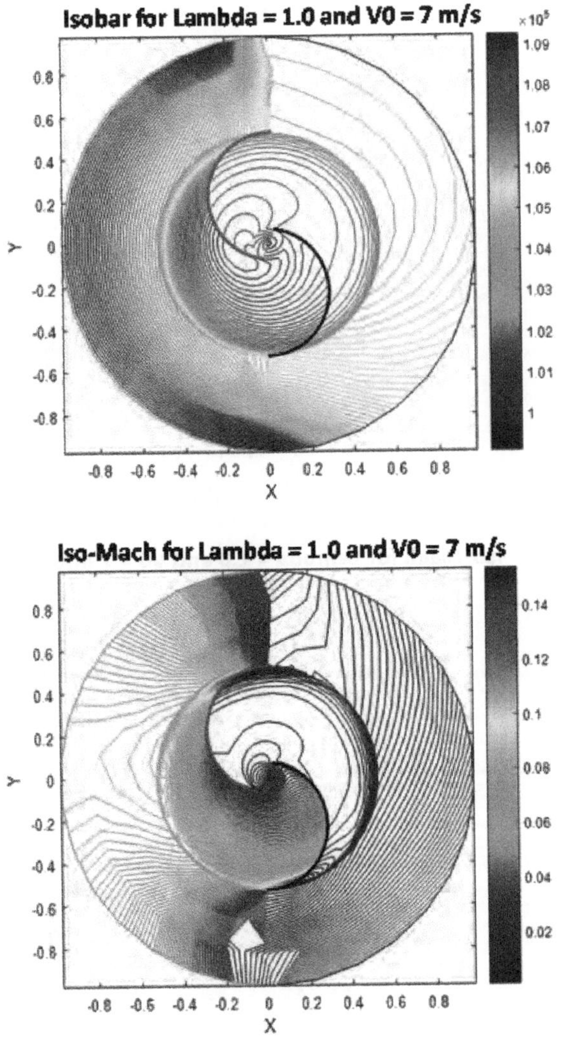

Figure 6.17. *Isobar curves at the instant t = 11.83 s, α = 90° after 13 rotations. For a color version of this figure, see www.iste.co.uk/prudhomme/fluid2.zip*

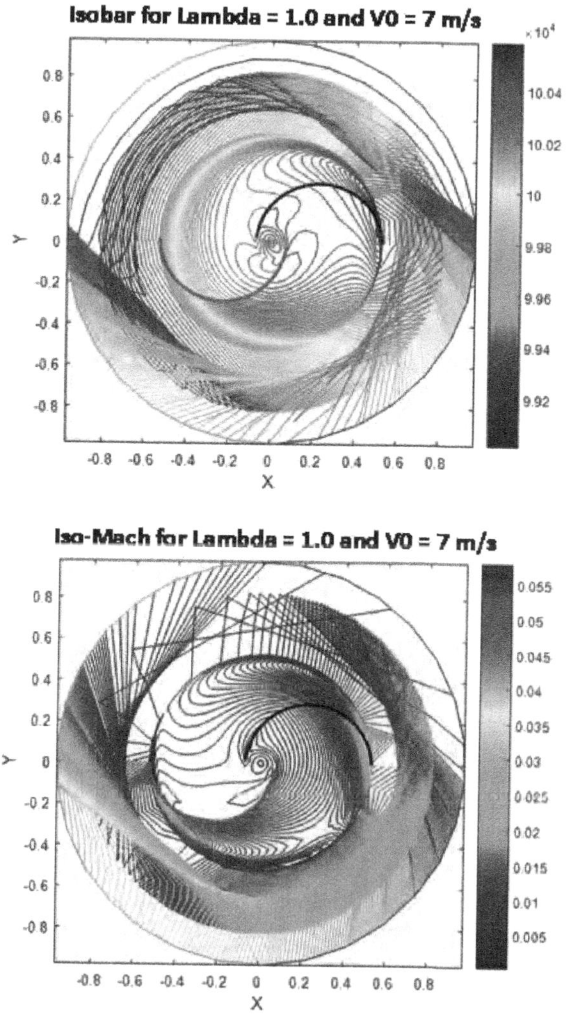

Figure 6.18. *Isobar curves at the instant $t = 11.94$ s, $\alpha = 180°$ after 13 more rotations. For a color version of this figure, see www.iste.co.uk/prudhomme/fluid2.zip*

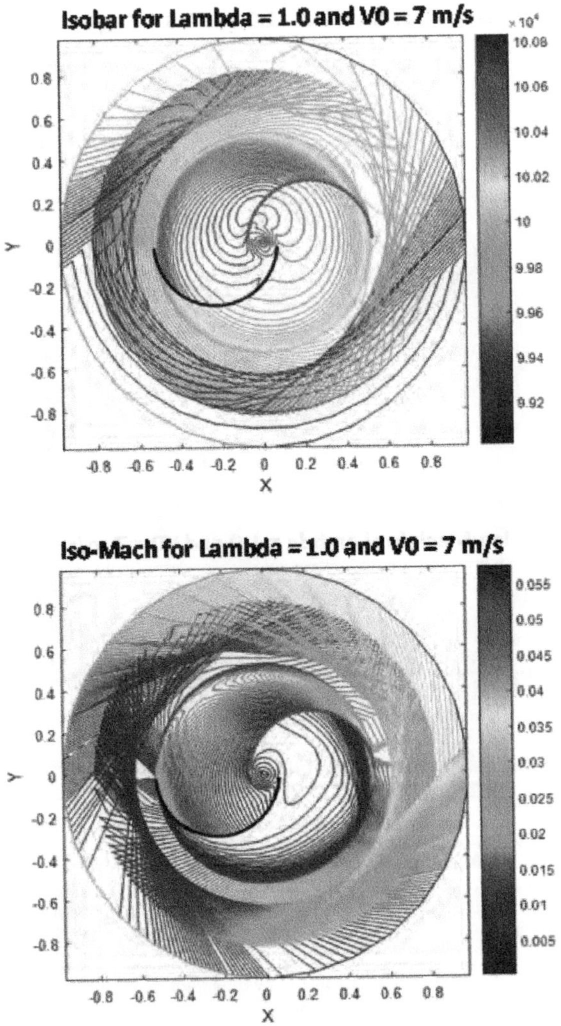

Figure 6.19. *Isobar curves at the instant t = 12.06 s, α = 270° after 13 more rotations. For a color version of this figure, see www.iste.co.uk/prudhomme/fluid2.zip*

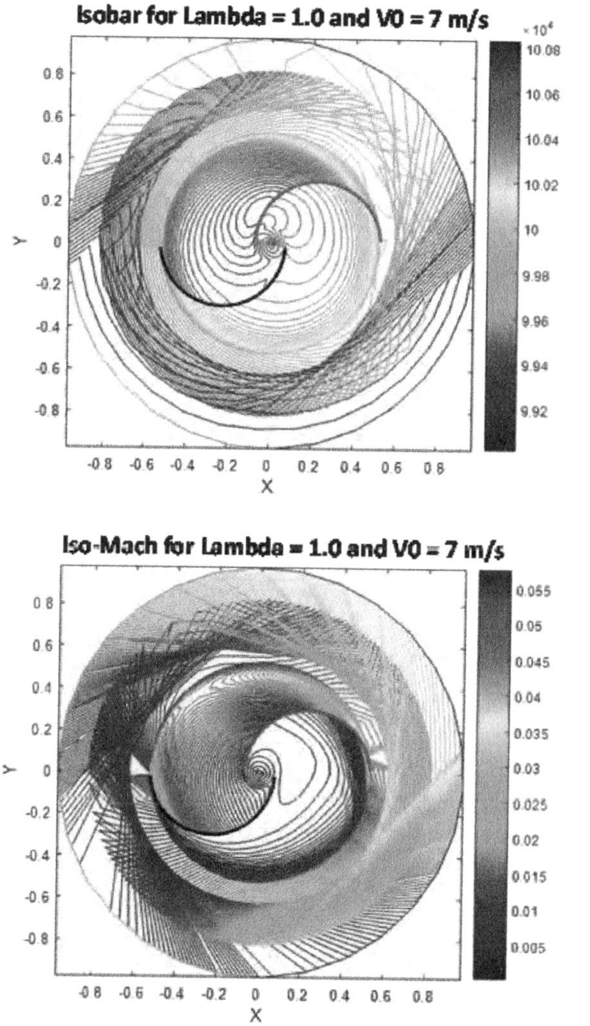

Figure 6.20. *Isobar curves at the instant t = 12.18 s, α = 360° after 13 more rotations. For a color version of this figure, see www.iste.co.uk/prudhomme/fluid2.zip*

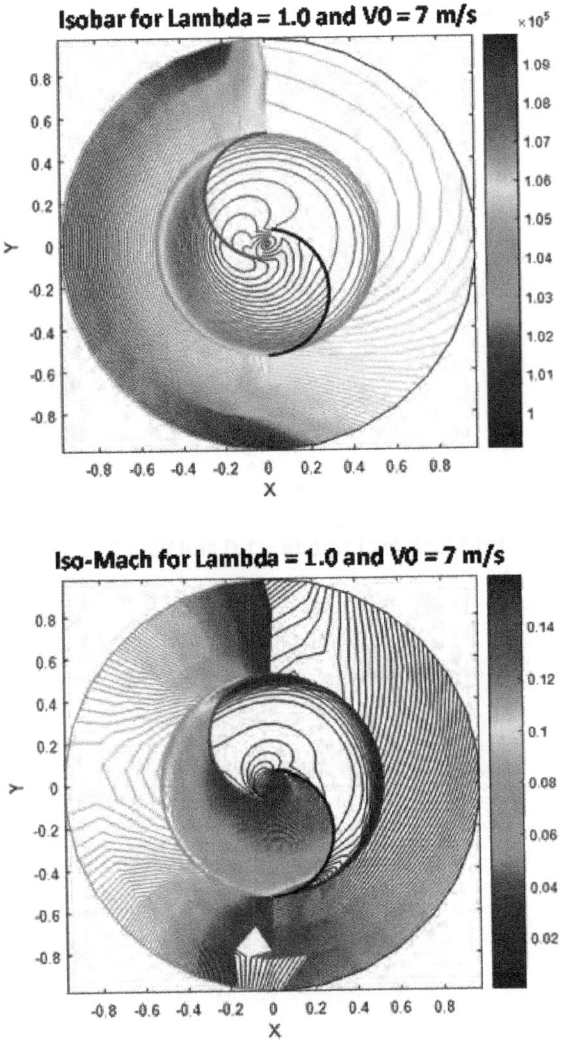

Figure 6.21. *Isobar curves at the instant t = 12.29 s, α = 90° after 14 rotations. For a color version of this figure, see www.iste.co.uk/prudhomme/fluid2.zip*

Figures 6.6–6.12 represent the isobar and iso-March curves for different values of the advance parameter λ, with an absolute velocity V = 5 m/s and a number of iterations IT = 43,000, during a period e t = 0.23 s. It has been noted that the advance parameter has a considerable influence on the rotation speed of a wind turbine. Its optimal value is about 0.9, but this partly depends on the geometric data of the turbine. Figures 6.13–6.16 show that the flow rate of fluids varies greatly or is perturbed when the power is on. The flow velocity of the fluid is almost transonic across the internal sweep of a rotor, thus creating a whirlwind, as shown in Figure 6.13.

Figures 6.13–6.21 show the same results as before, however the advance parameter and absolute velocity are fixed, while the number of iterations varies. The simulation results show that the physical phenomena that are encountered vary at each instant due to the non-stationary fluid flow. It is seen that the zone with the most perturbation is located at the level of the leading edges, trailing edges and, especially, in the internal overlap of the rotor. A depression was observed in a zone corresponding to the upper surface of the advancing blade (Figures 6.17–6.20), and an overpressure on the underside of the returning blade (blade shown in black). This pressure difference creates a torque that causes the wind turbine to turn. The turbulence in these two zones is slightly influenced by the acceleration of the particles, generated by the curvature of the blades and the rotation effect of the turbine. We also saw that the isobar curves correlated with the relative iso-Mach curves, i.e. in the high-pressure region, the fluid flow velocity is low. Bernoulli's principle is then verified, as the pressure is inversely proportional to velocity. In a non-stationary study, the system achieves stability if the results obtained between two successive time intervals are identical. This is proven by Figures 6.16, 6.17 and 6.21: the results obtained at the instant t = 11.83 s (13 rotations of a rotor) are identical to the results obtained at the instant t = 12.29 s (14 rotations of a rotor).

From six rotations of a turbine rotor onwards, the results obtained are similar. To obtain an accurate result, we need 12.29 s in real time, but four days in machine time. This duration varies depending on the memory capacity of the computers, as well as the boundary conditions fixed at the borders of the calculation domain.

6.6. Conclusion

The aim of this study was to analyze the aerodynamic behavior of fluids within a Savonius wind turbine rotor in order to improve it. This type of fluid is modeled by Euler equations for which analytical solutions are still only possible for simple structures with small degrees of freedom [KAN 18]. As a result, a numerical solution using the explicit finite difference method was adopted by using a scheme with artificial corrected viscosity order to minimize computation time without invalidating the results.

The simulation results showed that the flow of fluids around a profile in motion is perturbed due to the initial and boundary conditions. However, these perturbations are diminished during the calculation by applying a numerical viscosity. This simulation in the inter-blade channel in a conventional Savonius wind turbine reveals a depression downstream and overpressure upstream. This variation in pressure between the inlet and outlet of the calculation domain is inversely proportional to the variation in velocity. Consequently, the wind turbine decreases pressure, hence its name "turbine". The results of our calculations are highly reliable and effective. Nonetheless, an opposite approach will be appreciated in order to improve the geometry of the machine to obtain the required power.

6.7. Acknowledgments

The authors wish to thank all of the people who supported them in this research. They especially wish to thank the team at the MFSEA Laboratory (Laboratory of Applied Fluid Mechanics and Energetics Systems) at Antsiranana University (Madagascar) for their expertise.

6.8. References

[BUT 13] BUTAUD P., AURELIEN B., THIBAUD M. *et al.*, "Etude aérodynamique instationnaire d'un rotor éolien de Savonius : mise en évidence de l'influence de la portance", *Congrès mécanique*, hal-00904222, August 2013.

[DEB 13] DEBNATH P., RAJAT G., "Flow physics analysis of three-bucket helical Savonius rotor at 90-degree twist angle using CFD", *International Journal of Modern Engineering Research (IJMER)*, vol. 3, no. 2, pp. 739–746, March–April 2013.

[DOB 12] DOBREV I., MASSOUH F., "Exploring the flow around a Savonius Wind Turbine", Session 4.2: Rotating Machinery, *16th International Symposium on Applications of Laser Techniques to Fluid Mechanics*, Lisbon, Portugal, 9–12 July 2012.

[DWI 16] DWIYANTORO B.A., TRIYOGI Y., VIVIEN S., "Structural design optimization of vertical axis wind turbine type Darrieus-Savonius", *ARPN Journal of Engineering and Applied Sciences*, vol. 11, no. 2, January 2016.

[ELK 15] EL KHCHINE Y., SRITI M., OUARIAGLI M., "Simulation des performances aérodynamiques d'une éolienne à axe horizontal", *12th Mechanics Congress*, Casablanca, Morocco, 21–24 April 2015.

[GHA 14] GHANEGAONKAR P.M., KAWADE R.K., GARG S., "Conceptual model of vertical axis wind turbine and CFD analysis", *International Journal of Innovative Research in Advanced Engineering (IJIRAE)*, vol. 1, no. 3, pp. 76–81, May 2014.

[HAD 13] HADI ALI M., "Experimental comparison study for Savonius Wind Turbine of two and three blades at low wind speed", *International Journal of Modern Engineering Research (IJMER)*, vol. 3, no. 5, pp. 2978–2986, September–October 2013.

[KAN 13] KANG C., XIN Y., YULI W., "Turbulent flow characteristics and dynamics response of a vertical-axis spiral rotor", *Energies*, vol. 6, no. 6, pp. 2741–2758, available at: https://doi.org/10.3390/en6062741, 2013.

[KAN 18] KANYIKI T., "Simulation par la méthode des éléments finis du comportement vibratoire d'un rotor", *ISTE OpenScience*, London, UK, available at: http://openscience.fr/Simulation-by-the-Finite-Element-Method-of-the-Vibratory-behaviour-of-a-Rotor, 2018.

[MAR 17] MARCHEVSKY ILIA K., VALERIA V.P., "Numerical simulation of wind turbine rotors autorotation by using the modified LS-STAG immersed boundary method", *International Journal of Rotating Machinery*, vol. 2017, available at: https://doi.org/10.1155/2017/6418108, October 2017.

[NOU 15] NOUIOUA A., RABAH D.A., "Modélisation d'un écoulement pulsatif et oscillatoire autour d'une pale d'éolienne HAWT Application au décrochage dynamique", *Revue des énergies renouvelables*, vol. 18, no. 1, March 2015.

[PRU 18] PRUD'HOMME R., "Modélisation d'interfaces avec phase fluide", *ISTE OpenScience*, London, UK, available at: http://openscience.fr/Modelisation-d-interfaces-avec-phase-fluide, 2018.

[SON 16] SONG L., HONG-ZHAO L., ZONG-XIAO Y., "Orthogonal analysis based performance optimization for vertical axis wind turbine", *Mathematical Problems in Engineering*, vol. 2016, no. 6241360, available at: http://dx.doi.org/10.1155/2016/6241360, 2016.

[SOS 18] SOSSEY-ALAOUI I., RADI B., "Une étude aérodynamique d'une pale d'éolienne à axe horizontal", *ISTE OpenScience*, London, UK, available at: https://www.openscience.fr/IMG/ pdf/iste_incertfia18v2n1_4.pdf, 2018.

[WAH 13] WAHYUDI B., SUDJITO S., SLAMET W. *et al.*, "A simulation study of flow and pressure distribution patterns in and around of tandem blade rotor of Savonius (TBS) hydrokinetic turbine model", *Journal of Clean Energy Technologies*, vol. 1, no. 4, available at: http://www.jocet.org/index.php?m=content&c=index&a=show&catid=29&id=332, October 2013.

[WID 12] WIDODO W.S., CHIN A.C., HAERYIP S. *et al.*, "Design and analysis of 5 kw Savonius rotor blade", *Global Engineers & Technologists Review*, vol. 2, no. 8, 2012.

[ZHA 17] ZHANG B., BAOWEI S., ZHAOYONG M. *et al.*, "A novel parametric modeling method and optimal design for Savonius wind turbines", *Energies*, vol. 10, no. 301, available at: www.mdpi.com/journal/energies, March 2017.

List of Authors

Kwassi ANANI
Faculty of Sciences
Department of Mathematics
University of Lomé
Togo

Jean-Paul CALTAGIRONE
Arts et Metiers Institute of Technology
INRAE
I2M Bordeaux
Talence
France

Tarik CHAKKOUR
ISAE-ENSMA
Chasseneuil-du-Poitou
France

Christian CHAUVEAU
ICARE
CNRS
Orléans
France

Mahouton Norbert HOUNKONNOU
University of Abomey-Calavi
Republic of Benin

Roger PRUD'HOMME
Institut d'Alembert
Sorbonne University
CNRS
Paris
France

Francis RAVELOSON
MFSEA Laboratory
Université d'Antsiranana
Madagascar

Delphin TOMBORAVO
MFSEA Laboratory
Université d'Antsiranana
Madagascar

Stéphane VINCENT
University of Paris-Est
Marne-La-Vallée
France

Roger VONY
MFSEA Laboratory
Université d'Antsiranana
Madagascar

Index

Summary of Volume 1

Chapter 2. Simulations of Turbulent Two-Phase Flows with Phase Change Using a Multifield Approach Combined with LES

Solène GOUÉNARD, Stéphane VINCENT and Stéphane MIMOUNI

Chapter 3. An Original Approach to Extract Momentum and Heat Transfers from Particle-Resolved Simulations of Particulate Flows

Mohamed-Amine CHADIL, Stéphane VINCENT and Jean-Luc ESTIVALÈZES

Chapter 4. Interfaces and Critical Fluids

Roger PRUD'HOMME

Chapter 5. Shear-Induced Anomalies in the Brownian Motion of Particles in Strongly Fluctuating Near-Critical Mixtures

Daniel BEYSENS

Chapter 6. Basics on Interfaces in Combustion

Roger PRUD'HOMME

Other titles from

iSTE

in

Fluid Mechanics

2020

SAD CHEMLOUL Nord-Eddine
Dimensional Analysis and Similarity in Fluid Mechanics

2019

CALTAGIRONE Jean-Paul
Discrete Mechanics: Concepts and Applications

2018

ABABOU Rachid
Capillary Flows in Heterogeneous and Random Porous Media 1

2017

TARDU Sedat
Wall Turbulence Control

2015

CALTAGIRONE Jean-Paul
Discrete Mechanics

FOHR Jean-Paul
Heat and Moisture Transfer between Human Body and Environment

2014

PRUD'HOMME Roger
Flows and Chemical Reactions in Heterogeneous Mixtures
Flows and Chemical Reactions in an Electromagnetic Field

TARDU Sedat
Transport and Coherent Structures in Wall Turbulence

2013

BORGHI Roland, ANSELMET Fabien
Turbulent Multiphase Flows with Heat and Mass Transfer

DÉLERY Jean
Three-dimensional Separated Flow Topology: Critical Points, Separation Lines and Vortical Structures

PRUD'HOMME Roger
Flows and Chemical Reactions in Homogeneous Mixtures

2012

BOUTIER Alain
Laser Velocimetry in Fluid Mechanics

BOUTIER Alain
Laser Metrology in Fluid Mechanics: Granulometry, Temperature and Concentration Measurements

PRUD'HOMME Roger
Flows and Chemical Reactions

2011

MORY Mathieu
Fluid Mechanics for Chemical Engineering

TARDU Sedat
Statistical Approach to Wall Turbulence

2010

COLIN Stéphane
Microfluidics

DÉLERY Jean
Handbook of Compressible Aerodynamics

GUINOT Vincent
Waves Propagation in Fluids – 2nd edition

SOULI Mhamed, BENSON J. David
Arbitrary Lagrangian Eulerian and Fluid-structure Interaction

2009

FAVRE-MARINET Michel, TARDU Sedat
Convective Heat Transfer

2008

LALAUZE René
Physical Chemistry of Solid-Gas Interfaces

PEUBE Jean-Louis
Fundamentals of Fluid Mechanics and Transport Phenomena

SCHIESTEL Roland
Modeling and Simulation of Turbulent Flows

Printed and bound by CPI Group (UK) Ltd, Croydon, CR0 4YY
16/09/2026
14963076-0007